Communications in Medical and Care Compunetics

Volume 2

Series Editor
Lodewijk Bos, International Council on Medical and Care Compunetics,
Utrecht, The Netherlands

For further volumes:
http://www.springer.com/series/8754

This series is a publication of the International Council on Medical and Care Compunetics.

International Council on Medical and Care Compunetics (ICMCC) is an international foundation operating as the knowledge centre for medical and care compunetics (COMPUting and Networking, its EThICs and Social/societal implications), making information on medicine and care available to patients using compunetics as well as distributing information on use of compunetics in medicine and care to patients and professionals.

Lodewijk Bos · Denis Carroll
Luis Kun · Andrew Marsh
Laura M. Roa

Editors

Future Visions on Biomedicine and Bioinformatics 2

A Liber Amicorum in Memory of Swamy Laxminarayan

 Springer

Lodewijk Bos
International Council on Medical and
  Care Compunetics (ICMCC)
Stationsstraat 38
3511 Utrecht
The Netherlands
e-mail: lobos@icmcc.org

Denis Carroll
Head of KTP Unit
University of Westminster
Regent Street 309
London W1B 2UW
UK
e-mail: d.c.carroll@westminster.ac.uk

Luis Kun
Center for Information Assurance
  Education
National Defense University
300 5th Avenue SW, Marshall Hall
Washington, DC
20319-5066
USA
e-mail: kunl@ndu.edu

Andrew Marsh
VMW Solutions Ltd
Northlands Road, 9
Hampshire SO51 5RU
UK

Laura M. Roa
Departamento de Ingenería de Systemas y
  Automática
University of Seville
E.T.S. de Ingenieros Indus
Camino Descubrimientos s/n–Isla Cartuja
41092 Sevilla
Spain
e-mail: laura@esi.us.es

ISSN 2191-3811
ISBN 978-3-642-26963-9
DOI 10.1007/978-3-642-19554-9
Springer Heidelberg Dordrecht London New York

e-ISSN 2191-382X
ISBN 978-3-642-19554-9 (eBook)

Cover design: eStudio Calamar S.L.

Printed on acid-free paper

Springer is part of Springer Science+Business Media (www.springer.com)

Contents

Health Informatics: A Roadmap for Autism Knowledge Sharing

Ron Oberleitner, Rebecca Wurtz, Michael L. Popovich, Reno Fiedler, Tim Moncher, Swamy Laxminarayan and Uwe Reischl

Abstract With the prevalence of diagnosed autism on the rise, increased efforts are needed to support surveillance, research, and case management. Challenges to collect, analyze and share typical and unique patient information and observations are magnified by expanding provider caseloads, delays in treatment and patient office visits, and lack of sharable data. This paper outlines recommended principles and approaches for utilizing state-of-the-art information systems technology and population-based registries to facilitate collection, analysis, and reporting of autism patient data. Such a platform will increase treatment options and registry information to facilitate diagnosis, treatment and research of this disorder.

Keywords Autism · Patient information · Patient observations · Information systems technology · Population-based registries

First published in Bos L, Laxminarayan S, Marsh A, editors. Medical and Care Compunetics. IOS Press; 2005. p. 321–6. ISBN: 978-1-58603-520-4

R. Oberleitner (✉)
e-Merge Medical Technologies, Boise, ID, USA
e-mail: rono@talkautism.org

R. Wurtz · M. L. Popovich · R. Fiedler · T. Moncher
Scientific Technologies Corporation, Tucson, AZ, USA

S. Laxminarayan
Institute of Rural Health and Biomedical Research Institute,
Idaho State University, Pocatello, USA

U. Reischl
Center of Health Policy, Boise State University, Boise, USA

Commun Med Care Compunetics (2011) 2: 1–8
DOI: 10.1007/8754_2010_10
© Springer-Verlag Berlin Heidelberg 2011
Published Online: 29 March 2011

1 Background

Autism spectrum disorder (autism) is characterized by a range of neurological anomalies that typically include varying degrees of communication deficits and repetitive negative social behaviors. A tenfold increase in the incidence of autism over the past 15 years has been documented and is regarded as a significant public health concern. Despite the documented increase in the incidence of autism, the cause(s) of this disorder and appropriate treatment remain mysterious. The NIH road map emphasizes the need for developing phenotypic signatures based on available evidence including documentation of behavioral, clinical and genetic traits, as well as contributions by the basic sciences and applied bioengineering such as medical imaging outcomes, auditory phenomenology, neuroscience, and brain modeling studies.

Current population-based databases include a number of cross sectional studies sponsored by the CDC (Autism and Developmental Disabilities Monitoring Network [ADDM Net] and NIMH). These involve partnerships between a variety of governmental agencies, universities, and leading nonprofit organizations. Database initiatives that have been spearheaded include the Autism Genetic Resource Exchange, Autism Treatment Network, and Autism Tissue Program. Each of these offer contributions to the understanding of autism, but have significant limitations

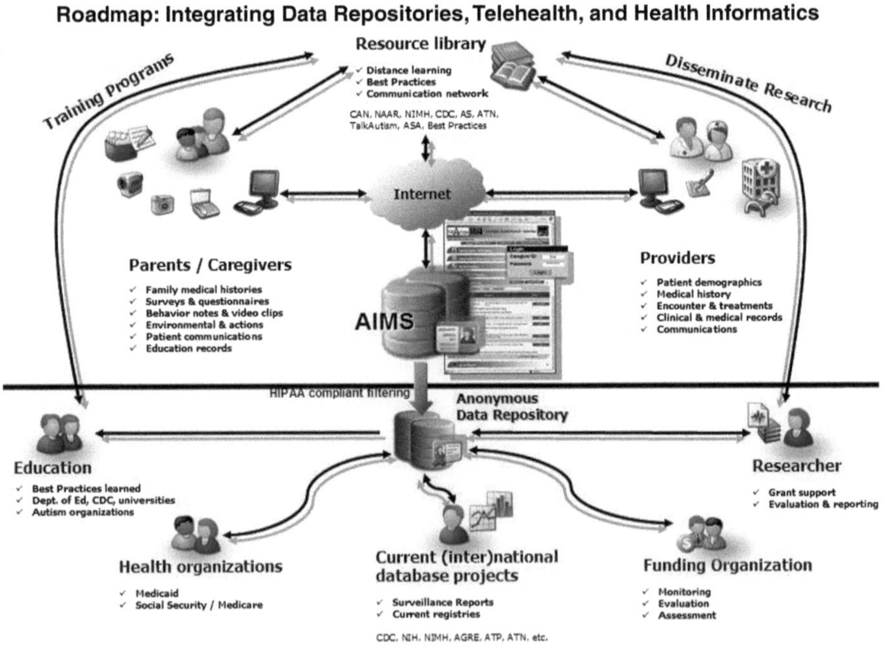

Fig. 1 Illustration of AIMS[TM] to service patients, parents, and healthcare providers, while supporting researchers, health organizations, and funding agencies in understanding more about autism spectrum disorders

in terms of ease of use, costs to build and maintain, and interoperability with other database projects.

In the National Institute of Mental Health's April 2004 *Congressional Appropriations Committee Report on the State of Autism Research* [1], the authors list the following obstacles, among others, to understanding the causes of and treatments for autism.

- Lack of a national autism twin registry that would allow researchers to access a large sample of well-defined twins where at least one twin is affected by autism.
- Lack of multi-site, high-risk population studies (i.e. pregnancies and infant siblings of individuals with autism) that would allow for increased knowledge about risk factors, early development of autism, and enhanced characterization of the disorder.
- Need for enhanced mechanisms to involve voluntary organizations, industries and potential donors in all stages of research design and implementation.

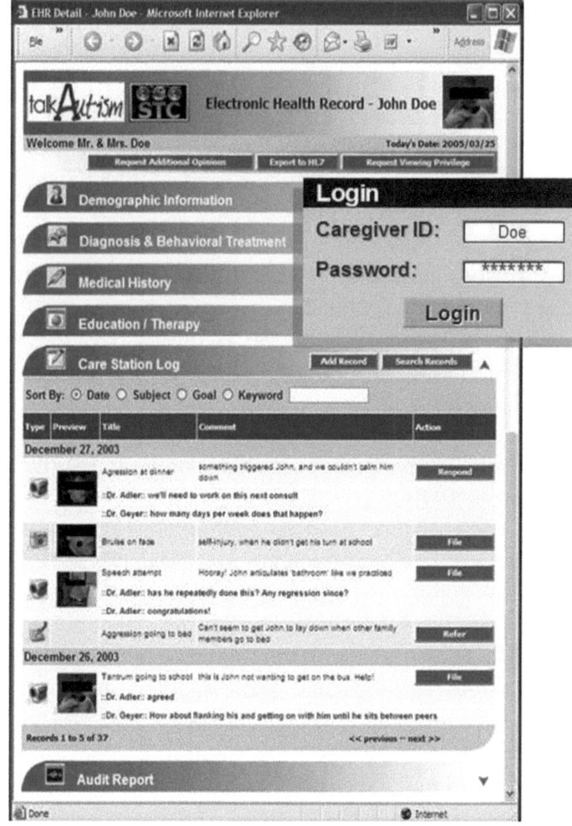

Fig. 2 Sample view of an autism EHR as developed by e-Merge/TalkAutism (Boise, ID) and STC (Tucson, AZ)

The shortcomings related to a lack of information resources can be overcome by the design and implementation of a longitudinal, person-based autism registry that would leverage the benefits provided by telehealth and the benefits offered by an interoperability infrastructure which integrates and builds on information already generated by the above-referenced initiatives. This paper outlines a vision for such a registry.

Complementary to necessary in-person examinations, the value of telemedicine and information technology to support the evaluation, diagnosis, and treatment of autism by the community of parents, health care providers, educators, and researchers has been outlined [2] (Fig. 1). To date, the ability to create a sharable information resource to support the diverse community of stakeholders is limited. The following illustration provides the concept for a new Autism Information Management System (AIMS). This system is designed, in part, to create a complementing patient registry that will be interoperable in relation to the current database initiatives, while providing a platform of sharable information to support the mission and goals of the various stakeholders.

1.1 Caregivers and Providers

The primary concept for the AIMS is a "Parent/Provider driven Person-Centric Information Environment" enabled by a web-based electronic health record (EHR), designed and maintained to enhance treatment options for caregivers. Caregivers would own the data and would have final jurisdiction in matters of access by providers. Providers (clinics, health professionals, therapists, specialized educators etc.) and caregivers (parents, other family members, paraprofessionals, respite workers, therapists, etc.) could complement in-person visits by communicating directly via a telehealth platform.

To help foster optimal use of this EHR, the system should incorporate an always-updating online portal resource library tailored to the caregivers and providers. Such a library will increase access to distance learning, updatable resource directories, and online communication forums involving other caregivers and health professionals is optimal to provide support and incentive to update the EHR.

An EHR can be used to capture and transmit patient behavior in a natural environment via input into text, and data capture devices like stethoscopes, or even cell phones and videophones. For example, images and video clips from a digital camera can send date linked to the treatment activities, milestones, or concerning behaviors. This can facilitate patient case management by providing visibility and insights into episodes that occur in their natural environment, and will allow a provider to remotely evaluate situations occurring at the moment of concern, and without delays or distractions found in a typical office environment. This type of system minimizes the impact on the individual with autism while maximizing the utilization of the provider. The system also offers the opportunity of the parent/guardian to record accurate information in a timely fashion, which is of utmost concern to most.

By providing such support and communication benefits, the platform is also a convenient medium for researchers to request voluntary information to facilitate research via surveys, questionnaires or with unique data capturing technology (Fig. 2). And as seen in other applications of telemedicine, there is savings realized by reduced travel for both professionals and families, comparable satisfaction to inperson visits, and advantages of accurate case documentation—all contributing to justify the technology hosting fee for this platform.

1.2 'Patient Case' to 'Anonymous Data' Repository

The design of the AIMS targets the need for researchers, health professionals, and educators to collect information about populations of individuals with autism. The vision is to allow anonymous data sets to be built based upon individual patient cases propagated in an individual's EHR, that can be integrated and coexisting with other database projects. De-identified information will be combined to create an extended knowledgebase to support applied research as well as information sharing of "best practices." Funding organizations would also be provided the ability to use the information to monitor and evaluate the impact of their service support.

Technical characteristics of this system would follow recent public health information development standards [3] and would build upon the lessons learned in developing population-based registries such as immunization information tracking systems [4]. Specifically, the system would exhibit the following features:

- Would utilize a secure web-based technology to support data collection and information retrieval in an easy-to-use format.
- The information database would be relational and person-centric to support individual case management, individual encounters, and would include treatment-based tracking.
- The system would include appropriate tools needed to capture and link video clips, family observations, and health histories related to time and space (i.e. environmental conditions).
- The system would include the necessary tools to support documentation, research, and reporting.

In order to achieve these goals, the AIMS must have the capability to electronically transfer information in a secure environment. The use of a Master Patient Index (MPI) to uniquely identify patients and to protect confidentiality will be essential. The underlying patient/provider database would contain defining data fields and code sets to support patient management including the following:

- Patient identification and demographics
- Family history
- Longitudinal medical history
- Epidemiologic questionnaires: i.e. exposures
- Time stamped behavior characteristics with attached video clips
- Treatment plans and parent progress reports
- Clinical and medical records

In addition to the core components, the system would allow attachment of added code sets such as:

- Co-morbidity (e.g. ADHD, sleep disorders, etc.)
- School records and reports
- Online treatment survey data
- Family observations of treatment efficacy

One of the essential design criteria will be to guard against information overload. In addition to the controls embedded in the data collection tools, it is recommended that "rule based" algorithms be employed to search for specific criteria, automating alerts for rapid provider notification and assessment.

2 Rationale

The typical health information system is one that is driven by patient encounters and maintained by providers or payers. These types of information systems currently do not support patient nor parent/guardian needs. They do not support research and reporting requirements. As such, additional information systems must be developed for clinical trials, patient registries, and statistical reporting. Resources are duplicated, additional costs incurred, and the ability to share lessons learned is curtailed or non-existent.

AIMS will be designed to collect information from diverse sources, store and share person-based case data and video, and monitor and report all value-added benefits. For example, there could be a module that can integrate school data in parallel. The ability to protect the privacy and confidentiality of individuals, providers, and research initiatives will require that information resources be limited to registered users and managed and controlled in compliance with HIPAA security standards.

The autism caregiver community should be especially motivated to adopt and propagate an accessible electronic health record that is easy to update and offers enhanced treatment for the affected individual(s) in their care. Many families maintain meticulous health history information because they typically visit multiple health providers and must therefore coordinate multiple stakeholders' understanding of their child's medical history. In schools, current best practices frequently require data collection and analysis to determine treatment effectiveness. Various technology options are appearing on the market to support families and educators in this regard.

There are a number of reasons why a patient-centric autism community telehealth platform is feasible at this time. National objectives have been established through current federal initiatives to facilitate the implementation of electronic health records (EHR). These initiatives require that health care information technology providers work with the community to establish standards for communication and data transfer. The relatively recent use of standard "case" definitions and data elements encourages the development of population-based databases for information sharing about population health indicators. This can directly lead to a better understanding of autism.

The national push towards more extensive use of electronic health records will encourage technology vendors to develop improved next-generation online health records systems. As more health data is created and stored electronically, there will be increasing opportunities to share information and more incentives to establish resources capable of recording longitudinal data on individuals. The impact of HIPAA to support patient confidentiality has also forced the information technology community to focus more on security and thus establish improved methodologies for protecting and sharing data.

In addition to national trends and standard implementation, there are recent examples of registries that have succeeded. Chronic disease and medical registry models including population-based immunization registries are being implemented and maintained by public health departments. These systems acquire data through the participation of both private and public health care providers. There are now technology, business practice and policy solutions available that capture patient demographics and health information electronically. These systems are also available through easy-to-use web-based applications and protect patient and provider confidently. These systems can be used as models for the implementation of autism-based registries.

3 Conclusion

A strong partnership between parents, providers, and teachers will be necessary to address the challenges of early diagnosis, treatment, and care of the children with autism. New telehealth technologies and electronic medical records storage and retrieval systems offer new opportunities for parents, providers and researchers to communicate their observations and findings to each other. We recommend the development of a new AIMS that will create a complementing patient registry that is interoperable in relation to current database initiatives while providing a platform of sharable information to support the mission and goals of parents, health care providers, teachers, and researchers involved with the autism spectrum disorder.

References

1. Congressional Appropriations Committee Report on the State of Autism Research. http://www. nimh.nih.gov/autismiacc/CongApprCommRep.pdf. Accessed Apr 2004.
2. Oberleitner R (2004). Talking to the autism community [interviewed by Semahat S. Demir]. IEEE Eng Med Biol Mag. 2005 Jan–Feb;24(1):14–15, 19.
3. CDC Publication, Public Health Information Network Functions and Specifications. http://www. cdc.gov/phin. Accessed Dec 2002.
4. Scientific Technologies Corporation, Immunization Information Systems, A Resource Guide. 2002 (a collection of white papers)

Bibliography

1. Oberleitner R, Laxminarayan S. Information technology and behavioral medicine: impact on autism treatment and research. In: Studies in healthcare technology and informatics, 2004. Vol 103, p. 215–22.
2. Oberleitner R, Laxminarayan S, Suri J, Harrington J, Bradstreet J. The potential of a store and forward tele-behavioral platform for effective treatment and research of autism. In: IEEE Eng Med Biol Soc, EMBC 2004. Conference Proceedings. Vol 2, p. 3294–6.

Non-telephone Healthcare: The Role of 4G and Emerging Mobile Systems for Future m-Health Systems

R. Istepanian, N. Philip, X. H. Wang and S. Laxminarayan

Abstract The next generation of "m-health technologies" is a new and evolving topic in the areas of telemedical and telecare systems. These technologies involve the exploitation of mobile telecommunication and multimedia technologies to provide better access to healthcare personnel on the move, by removing the key disadvantage of trailing wires in current systems. These technologies provide equal access to medical information and expert care by overcoming the boundaries of separation that exist today between different users of such medical information. A great benefit to all users will be a more efficient use of resources and far greater location independence. In this paper we will address some notes and future trends in these emerging areas and their applications for m-health systems. Especially we will discuss the role of 4G and emerging mobile systems for future m-health systems. The new technologies can make the remote medical monitoring, consulting, and health care more flexible and convenient. But, there are challenges for successful wireless telemedicine, which are addressed in this paper.

Keywords mHealth · 4G · Mobile Telemedicine Systems

First published in Bos L, Laxminarayan S, Marsh A, editors. Medical and care compunetics. IOS Press; 2005. p. 1–4. ISBN: 978-1-58603-520-4.

R. Istepanian (✉) · N. Philip · X. H. Wang · S. Laxminarayan
Mobile Information and Network Technologies Research Centre, School of Computing and Information Systems, Kingston University, Kingston upon Thames, Surrey, KT1 2EE, UK
e-mail: r.istepanian@kingston.ac.uk

S. Laxminarayan
Biomedical Information Engineering, Institute of Rural Health, Idaho State University, 921 South 8th Street, Gravely Hall, Pocatello ID 83209, USA

Commun Med Care Compunetics (2011) 2: 9–16
DOI: 10.1007/8754_2010_16
© Springer-Verlag Berlin Heidelberg 2011
Published Online: 26 March 2011

1 Introduction

Time and space constitute barriers between health-care providers and their patients and among health-care providers. Patients in rural areas, on a space shuttle flight, at accident scenes, en route to a hospital, in a submarine, etc., are often physically remote to appropriate care providers.

Telecommunication technologies have presented themselves as a powerful tool to break the barriers of time and space. With the introduction of high-bandwidth, digital communication technologies, it is possible to deliver audio, video, and waveform data to wherever and whenever needed.

The health-care industry may be poised to adopt wireless devices and applications in large numbers. Wireless technology may provide improved data accuracy, reduce errors, and result in overall improvement of patient care. The number of wireless devices in health-care is expected to triple by 2005, according to a study by Technology Assessment Associates. Wireless-enabled handheld usage by U.S. physicians is likely to climb to 55% by 2005, up from the current 18% [1].

The benefits of the wireless technology can be illustrated in a number of different examples [2]. Patient information can be obtained by health-care professionals from any given location because they can be connected wirelessly to the institution's information system. Physicians' access to patient histories, lab results, pharmaceutical information, insurance information, and medical resources would be enhanced, thereby improving the quality of patient care. Handheld devices can also be used in home health-care, for example, to fight diabetes through effective monitoring.

The major step from second- to third-generation and further to fourth-generation and beyond mobile communications was the ability to support advanced and wideband multimedia services, including email, file transfers, and distribution services, including email, file transfers, and distribution services like radio, TV, and software provisioning (e.g., software download). In general the combination and convergence of the different worlds of information technology (IT), media, and telecommunications will integrate communications. As a result mobile communication together with IT will penetrate various fields of society and especially telemedicine.

4G is expected to support faster- and larger-capacity transmissions, in order to provide high-resolution video and other applications seamlessly in a mobile environment [3].

Mobile telemedicine is a new and evolving area of telemedicine that exploits the recent development in mobile networks for telemedicine applications [4]. It was suggested that the next step in the evolution of telemedicine would be mobile telemedicine systems [5].

2 Overview of Mobile Telemedicine Systems

2.1 Current Mobile Technologies

In recent years there has been increased research on wireless telemedicine using current mobile communication systems, especially in USA and Europe, for conventional civilian and military use [6–14]. However, the increased equipment cost (such as satellite-based systems) and the limited bandwidth of the current generation of cellular telecommunication systems, have restricted the wider use of these systems within the most promising segments of the health care structures in general. However, in recent years some emerging 2.5G- and 3G-based m-health systems with Bluetooth medical wireless technologies have been cited in the literature [2].

2.2 Limitations of Existing Wireless Technologies for m-Health

The current mobile telemedicine systems can be characterised by the following drawbacks:

- The lack of a flexible and integrated telemedical linkage of the different mobile telecommunication options. This lack of linkage exists due to the difficulty of achieving operational compatibility between the telecommunication services and the current mobile standards.
- The high cost of communication links, especially between satellites and global mobile devices.
- The limited data transfer rate of the current mobile telephonic systems (around 9.6 Kbit/s). Specially when compared to the costly new ISDN 1 and Primary Rate Interface (PRI) of less than 2 Mbit/s, or even DSL at 8 Mbit/s [11].
- The limited availability of mobile internet connectivity and information access due to the current bandwidth limitations.
- Healthcare is a very complex industry and difficult to change.
- Organisational changes are very often required for healthcare institutions to benefit from mobile telemedicine.
- Those required organisational changes most likely have an impact on how physicians and other staff members lose or gain power as a result of those changes.
- The short-term and long-term economic consequences and working conditions for physicians and healthcare systems are not yet fully understood.
- The methods of payment for such mobile telemedicine are not yet fully developed and standardised.
- There is a lack of incentive for busy specialists to practice mobile telemedicine because it is seen as yet another imposition for which they are not paid.

- The currently available telemedicine equipment can sometimes be difficult to handle.
- There is a lack of integration between mobile telemedicine systems and other information systems e.g. referral and ordering systems, medical records etc.
- There are not enough numbers of demonstration projects that show mobile telemedicine's real savings potential.

The above are some of the factors that have hindered the wider applications of mobile telemedicine technologies thus far across health-care systems and on critical medical applications.

3 4G Technology

It is expected that the 4G mobile system will focus on seamlessly integrating the existing wireless technologies including GSM, wireless LAN, Bluetooth, and other newly developed wireless systems. So 4G system benefits from all those wireless technologies, as that currently there is no single system that is good enough to replace all the other technologies. Some key features of 4G networks are stated as follows:

1. High usability. 4G networks are all IP based heterogeneous networks that allow users to use any system at anytime and anywhere. Users carrying an integrated terminal can use a wide range of applications provided by multiple wireless networks.
2. Support for multimedia services at low transmission cost. 4G systems provide multimedia services with high data rate, good reliability and at low per-bit transmission cost.
3. This new-generation network will provide personalised service, in order to meet the demands of different users for different services.
4. 4G systems also provide facilities for integrating services. Users can use multiple services from any service provider at the same time.

The main technological characteristics of 4G systems are as follows:

1. The transmission speed is higher than 3G (min 50–100 Mb/s, average 200 Mb/s).
2. The system capacity is larger than 3G by 10 times.
3. The transmission cost per bit is decreased to 1/10 to 1/100 of 3G.
4. It should support internet protocols (IPv6).
5. 4G should have various qualities of services in order to provide many kinds of best effort multimedia services corresponding to users' demand.
6. User friendly services provider, as that user can access to so many services in short time as compared to other wireless systems that encounters long time waiting for response.

The future 4G service can provide communication with realistic sensation, in which 3D sound, light, and pressure fields are sent to another party to reproduce a situation. Therefore, virtual reality can be generated, letting you experience things as if you are "actually there" with bit rate of 50 Mb/s [14]. While via 3G system only the voice can be transmitted with any 2D image as the realistic sensation of the transmitted place with bit rate of 3.4 kb/s [14]. The current trend of research in field of wireless technology is towards building technology based human feelings (human communication). Human communications involves conveying feelings to communicate smoothly. Although videophones featuring images and virtual reality have accomplished visual communication of the user's appearance and the senses of virtual movement of the user environment, they alone are not enough to help convey feelings. So by adding voice, image, or data, and import the real physical sensations that complement feelings, the atmosphere around the user, and his/her physical movements in communication, it should be possible to establish a more sympathetic physical communication style. Such a communication style may be applied to a wide range of fields, including remote medical care.

4G advances will provide both mobile patients and normal working end users the choices that will fit their lifestyle and make easier for them to interactively get the medical attention and advice they need. When and where is required and how they want it regardless of any geographical barriers or mobility constraints. The concept of including high-speed data and other services integrated with voice services is emerging as one of the main points of the future telecommunication and multimedia priorities with the relevant benefits to citizen centered healthcare systems. These creative methodologies will support the development of new and effective medical care delivery systems into the 21-century. The new wireless technologies will allow both physicians and patients to roam freely, while maintaining access to critical medical information.

4 Next Generation m-Health Systems

The next few years will witness a rapid deployment in both wireless technologies and mobile internet based m-health systems with pervasive computing technologies. The increasing data traffic and demands from different medical applications and roaming application will be compatible with the data rates of 3G systems in specific mobility conditions. The implementation and penetration of 4G systems is expected to help close the gap in medical care. Specifically, in a society penetrated by 4G systems, home medical care and remote diagnosis will become common, check-up by specialists and prescription of drugs will be enabled at home and in underpopulated areas based on high-resolution image transmission technologies and remote surgery, and virtual hospitals with no resident doctors will be realised. Preventive medical care will also be emphasised: for individual health management, data will constantly be transmitted to the hospital through a built-in sensor in the individual's watch, accessories, or other items worn daily, and diagnosis results will be fed back to the individual. However, it is well known that current Healthcare systems are stuck with the equation:

Current organisation + New technology = Expensive current organisation.

Hence, the expectations are for these new-generation mobile and wireless technologies to be acceptable for sort of examples that represent challenges for these technologies such as:

1. Citizens become customers
2. Input measures are replaced by output measures
3. Citizen relationship costs fall
4. Taxes are lowered because of competition

In addition there is hope for the wider deployment of mobile telemedicine system because of some global changes, which are likely to have a major effect on the health-care industry. Those changes are:

- Increasing numbers of older adults and fewer young people so that to sustain the economy, the elderly will have to be persuaded to continue working longer. To be able to do this, a greater emphasis on the health of the elderly will mean an increase in demand for healthcare. At the moment an obstacle to the implementation of telemedicine is that commercial organisations do not regard the health economy as large enough to invest time and research. The growing demand for healthcare services and the reduced supply of service providers and caregivers will mean that telemedicine suddenly acquires a heightened importance.
- Fragmentation of care caused by the twin pulls of generalisation to push down costs and specialisation to meet the increasing needs of rapid advances. Co-operation in health-care, which has been anathema to healthcare workers, will have to be achieved by patient power rather than central directive.
- Increased patient expectation because of easier access to information will mean that the pre-eminence of the physician will be challenged. Patient lifestyles will mean that at least affluent ones will demand treatment wherever they are at the time because of a new leisure-oriented lifestyle. On the other hand patients at the lower end of the socio-economic scale may have to settle for lower expectations.
- Increased complexity of assessment, diagnosis, investigations and treatment will mean a knowledge explosion and the falling short of the quick dissemination of the knowledge and expertise. Again, telemedicine may serve a useful function of rapid dissemination of the skills and knowledge.

5 Cultural, Commercial and Operational Change

A nation's health service is fashioned by its economy, demography, culture, and medical tradition, among other factors. This identity poses a challenge to telemedicine, which can make it better. In addition it has to deal with the problem of component management. Component management derives from the observation that the providers and payers of health-care view health challenges only through the specific window of care for which they are responsible. One of the main incentives of health-care is the reimbursement, which is basic to the cost of health-care. Providers are forced to organise their packages into reimbursable ones. Any task, which falls outside these packages, tends to be overlooked or receive low priority. Thus component management systems serve patients poorly. Thus the emphasis is on treatment rather than prevention, there is a lack of incentive for providers to treat the entire disease process, which leads to an uncoordinated delivery system. Some other key factors that may accelerate the diffusion of m-health systems are:

1. Management perspective when planning the implementation of telemedicine to favour mobile solutions rather than fixed ones.
2. Economic perspective—mobile telemedicine costs and savings will probably appear on different accounts.
3. Development of payment systems to include mobile telemedicine.
4. Government intervention to fund exemplars of mobile telemedicine integrated into the health-care system.
5. Comprehensive assessments rather than feasibility reports by enthusiasts.
6. A need to educate and inform key players of what is available and what can be achieved in the future.
7. Clarification of the legal and ethical issues.

6 Conclusions

This paper addresses some of the fundamental issues and future scenarios regarding the next generation of mobile telemedicine systems. It is conclusive that some of the current and successful telemedicine systems will be more geared toward emerging wireless solutions in health-care scenarios that are not feasible with the current generation of cellular telephonic and internet services. The imminent launch of the next generation of wireless and internet technologies will fundamentally change the current structures of telemedical and healthcare delivery systems.

We can conclude that the main characteristics of the future 4G are: high communication speed, high capacity, low bit cost and IP-based technology. This promising technology will play a very important roll in telemedicine applications.

References

1. Pattichis CS, Kyriacou E, Voskarides S, Istepanian RSH. Wireless telemedicine systems: an overview. IEEE Antennas and Propagation. 2002;44(2):143–53.
2. Yao, Wenbing, Istepanian, RSH. '3G Mobile Communications for Wireless Tele-Echography Robotic System', Proceedings of the 6th World Multiconference on Systemics, Cybernetics and Informatis-SCI2002 Conference XV: Mobile/Wireless Computing and Communications

Systems III, Ed. by Callaos N, Duale A, Benedicenti L. Orlando, Florida, USA. p. 138–42, 14–18 July 2002.
3. Hui SY, Yeung KH. Challenges in the migration to 4G mobile systems. IEEE Commun Mag. Dec 2003:54–9.
4. Tachakra S, Wang XH, Istepanian RSH, Song YH. Mobile e-Health: the unwired evolution of telemedicine. Telemed J e-Health. 2003;9(3):247–57.
5. Istepanian RSH, Kyriacou E, Pavlopoulos S, Koutsouris D. Wavelet compression methodologies for efficient medical data transmission in wireless telemedicine system. J Telemed Telecare. 2001;7 Supp 1:14–6.
6. Istepanian RSH, Woodward B. Programmable underwater acoustic telemedicine system. Acoustica. 2002.
7. Istepanian RSH, Laxminaryan S. UNWIRED, the next generation of wireless and internetable telemedicine systems- editorial paper. IEEE Trans Inf Technol Biomed. Sept 2000;4(3):189–94.
8. Istepanian RSH, Pertrossian A. Optimal wavelet-based ECG data compression for mobile telecardiology system. IEEE Trans Inf Technol Biomed. Sep 2000;4(3):189–94.
9. Istepanian RSH, Nikogossian HA. Telemedicine in Armenia: a perception of telehealth services in the former Soviet Republics. J Telemed Telecare. 2000;6:268–72.
10. Istepanian RSH. Telemedicine in the United Kingdom, current status and future prospects. IEEE Trans Inf Technol Biomed. 1999;3(1):158–9.
11. Istepanian RSH, Petrosian AA. Wavelet zonal coding for ECG data compression. Med Biol Eng Comput. 1999;37 Suppl 1:369–70.
12. Richards C, Woodward B, Istepanian RSH. Exploiting mobile telephone technology for telemedicine applications. Med Biol Eng Comput. 1999;37 Suppl 1:110–1.
13. Istepanian RSH, Woodward B, Balos P, Chen S, Luk B. The comparative performance of mobile telemedical systems using the IS-54 and GSM cellular telephone standards. J Telemed Telecare. 1999;5(2):97–104.
14. Tachikawa K. A perspective on the evolution of mobile communications. IEEE Commun Mag. Oct 2003;66–73.

Bibliography

1. Woodward B, Istepanian RSH, Richard C. Design of a telemedical system using a mobile phone. IEEE Trans Inf Technol Biomed. 2001;5(1):13–5.
2. Istepanian RSH, Hadjileontiadis L, Panas S. ECG data compression using wavelets and higher order statistics methods for telemedical applications. IEEE Trans Inf Technol Biomed. 2001;5(2):108–15.

Understanding the Social Implications of ICT in Medicine and Health: the Role of Professional Societies

Brian M. O'Connell and Swamy Laxminarayan

Abstract In past times, engineers and other ICT professionals could normally function exclusively within an environment of purely technical dimensions. This sphere could be easily delineated from those involving policy, political or social questions. Consequently, these professions could well be characterized as generally isolated from mainstream society, engendering a condition that Zussman [1985] has described as a "technical rationality that is the engineer's stock-in-trade requir[ing] the calculation of means for the realization of given ends. But it requir[ing] no broad insight into those ends or their consequences". This condition has often led to a perceived technical mindset that according to Florman [1976], draws upon "the comfort that comes with the total absorption in a mechanical environment. The world becomes reduced and manageable, controlled and unchaotic".

Keywords Social implications · ICT · Professional societies

In a relatively short period of time, ICT has been radically transformed in both its capabilities and reach. Specifically, within the context of this event, the permeation of digital technologies into nearly every aspect of bioengineering and healthcare delivery have broken down the borders between technological pursuits and the larger dynamics of society. This has in turn has produced, according to Williams

Both the authors, Brian M. O'Connell and Swamy Laxminarayan, are deceased.
First published in: Bos L, Laxminarayan S, Marsh A, Editors. Medical and Care Compunetics. IOSPress; 2005. p. 5–7. ISBN 978-1-58603-520-4

B. M. O'Connell (⊠)
IEEE Society on Social Implications of Technology, Department of Computer Science, Central Connecticut State University, New Britain, CT 06050, USA

S. Laxminarayan
Biomedical Information Engineering, Idaho State University, Pocatello, ID 83209, USA

Commun Med Care Compunetics (2011) 2: 17–20
DOI: 10.1007/8754_2010_15
© Springer-Verlag Berlin Heidelberg 2011
Published Online: 29 January 2011

[6] a discipline that has "evolved into an open-ended Profession of Everything in a world where technology shades into science, into art, and into management, with no strong institutions to define an overarching mission". Within ICT, von Baeyer [1] affirms this status in noting "the frustration of engineers who have at their disposal a variety of methods for measuring the amount of information in a message, but to none deal with its meaning".

The cybernetics pioneer, Norbert Wiener [5] presaged the current climate when he wrote that "as engineering technique becomes more and more able to achieve human purposes, it must become more and more accustomed to formulate human purposes". This observation is particularly relevant to the global challenges presented within the context of e-Health as characterized by the Commission of the European Communities [2].

The development of medical technologies in the coming decades will make an ever greater impact on health services. Important innovations include the use of computers and robotics, the application of communications and information technology, new diagnostic techniques, genetic engineering, cloning, the production of new classes of pharmaceuticals, and the work now beginning on growing replacement tissues and organs. These developments can contribute significantly to improved health status.

The massive nature of the challenge is evidenced by a recent report of the Commission [3] which notes that:

- increased networking, exchange of experiences and data, and benchmarking, is also
- necessary at the European level in the health sector. Drivers for this include the need for
- improvements in efficiency, and the increased mobility of patients and health professionals
- under an emerging internal market in services. The situation requires the integration of
- clinical, organizational, and economic information across health care facilities, so as to
- facilitate virtual enterprises at the level of jurisdictions and beyond.

As predicted by Wiener and Williams, the far-reaching implications of these advances cannot be confined to infrastructure alone, and are certain to impact contemporary societal norms. It is notable that at the onset of its initiative, the Commission report [2] refers to the "significant ethical issues raised" in the process of developing new technologies. Viable responses to these challenges will not result from unilateral or detached applications of expertise. Instead they will require innovative approaches that reflect the present convergence of the technical and the social. Of foremost concern will be the establishment of a working dialogue among those in technological, legal, social and philosophical fields. Although such interactions have occurred in the past, the present need is arguably unique in history as it requires a dynamic and permanent partnership that is typified by more than superficial familiarity with other, often unfamiliar disciplines.

1 Diversity in Biology and Medicine

The diversity in biology and medicine has grown beyond belief especially with the introduction of advancing technologies. With diversity comes controversies, raising a whole gamut of ethical, legal, social, and/or policy issues. Typical examples include genetic engineering and biotechnology. Health care is a very sensitive area that requires individual protection against the invariable consequences of the social issues. As scientists and engineers, we have ambitious plans for ourselves. For example, as Francis Collins of the National Human Genome Research, has predicted (TIME, 2003), "I think it is safe to say we will have individualized, preventive medical care based on our own predicted risk of disease as assessed by looking at our DNA. By then each of us will have had our genomes sequenced because it will cost less than $100 to do that. And this information will be part of our medical record. Because we will still get sick, we will still need drugs, but these will be tailored to our individual needs. They will be based on a new breed of designer drugs with very high efficacy and very low toxicity, many of them predicted by computer models." These plans are already in action in ways that have triggered a whole series of social, ethical and policy issues associated with genetic and genomic knowledge and technology. No single institution can address on its own the various issues that are in interplay. Professional societies have a commitment to serve as an information base and provide the synergies required to bring together the interdisciplinary stakeholders to become involved in the debates.

2 SSIT as a Model

While formal institutional paradigms for this new mode of interaction are understandably sparse, the 33-year history of the Society on Social Implication of Technology (SSIT) of the Institute of Electrical and Electronic Engineers (IEEE) provides a useful model to explore interdisciplinary efforts. The SSIT consists of approximately 2,000 members worldwide. The scope of the Society's interests includes such issues as engineering ethics and professional responsibility; the use of technical expertise in public policy decision making; environmental, health and safety implications of technology and social issues related to energy, information technology and telecommunications. Throughout its existence, the SSIT has attracted a diverse membership consisting of engineers in academe and industry, computer scientists, educational specialists, attorneys, academic ethicists, philosophers, librarians, historians and other scholars and practitioners working in the humanities, the sciences and technology. The unique nature of SSIT is evidenced in the collaborative efforts of its members. Experience and knowledge are shared across disciplinary boundaries, making it possible to construct comprehensive pictures of socio-technical issues as well as strategies toward resolution of conflicts.

3 Conclusions

This presentation will consider the model of SSIT and those of other global professional societies in an effort to investigate the elements of successful collaboration within the context of ICT issues. It will further examine the dynamics that lead to open and fruitful dialogues across the disciplines.

References

1. von Baeyer C. Information: the new language of science. Cambridge: Harvard University Press; 2003.
2. Commission of the European Communities. Communication from the Commission to the Council, the European Parliament, the Economic and Social Committee and the Committee of the Regions on the Health Strategy of the European Community, Brussels. http://eur-lex.europa.eu/LexUriServ/LexUriServ.do?uri=COM:2000:0285:FIN:EN:.pdf. Accessed 16 May 2000.
3. Commission of the European Communities. Communication from the Commission to the Council, the European Parliament, the Economic and Social Committee and the Committee of the Regions on e-Health-making healthcare better for European citizens: an action plan for a European e-Health Area, Brussels. http://eur-lex.europa.eu/LexUriServ/LexUriServ.do?uri=COM:2004:0356:FIN:EN:.pdf. Accessed 16 April 2004.
4. Florman S. The existential pleasures of engineering. New York: St. Martin's Press; 1976.
5. Wiener N. God and golem, Inc. Cambridge: MIT Press; 1964.
6. Williams R. Retooling: a historian confronts technological change. Cambridge: MIT Press; 2002.
7. Zussman R. Mechanics of the middle class: work and politics among American engineers. Berkeley: University of California Press; 1985.

ICMCC the Information Paradigm

Lodewijk Bos, Swamy Laxminarayan and Andy Marsh

Abstract Below is the first published declaration of the goals of the ICMCC Foundation as well as an introduction to the ICMCC 2005 event, of which this article was the introduction to the proceedings.

Keywords Compunetics · Health Information Technology · Knowledge Centre · Dissemination · Health Information

1 Introduction

Business-to-business (B2B) and business-to-customer (B2C) approaches have been considered to be sound practices in the application of ICT (Information and Communication Technology) in commerce and industry.

In the medical and care areas, these concepts have not yet been common practice. But with the enormous explosion of heterogeneous information modalities in health care, the need for applying such concepts is obvious. However despite the limited research done so far in evaluating the possible effects, it is to be expected, that these practices will bring forth significant benefits to both the medical and care professionals and the consumer/patients.

First published in Bos L, Laxminarayan S, Marsh A, editors. Medical and care compunetics. IOS Press; 2005. p. 1–4. ISBN: 978-1-58603-520-4.

L. Bos (✉), S. Laxminarayan and A. Marsh
Members of the Board, ICMCC Council, Utrecht, The Netherlands
e-mail: lobos@icmcc.org

Commun Med Care Compunetics (2011) 2: 21–25
DOI: 10.1007/8754_2010_14
© Springer-Verlag Berlin Heidelberg 2011
Published Online: 9 January 2011

2 ICMCC 2004, the History

In September 2004 the International Council on Medical and Care Compunetics (ICMCC) was founded to create the infrastructure necessary for the B2B and B2C concepts in the medical and care domains. The creation of the council was a logical consequence of the first Congress on Medical and Care Compunetics held in the Hague, in June 2004 [1].

New and innovative in its format, the Congress was an off-shoot of ideas that were put together in April 2003 to emphasize the computing and networking synergies in medicine and (health) care. The term compunetics was coined to represent the union of the latter. Contrary to the traditional sessions-oriented conferences, ICMCC represented a meeting created around a cluster of special workshops in closely interrelated areas of compunetics. The call for workshops resulted in 18 workshops of either half a day or a full day. People from all over the world including Europe, USA, South America, and Israel participated in the workshops. Conference participants came from 26 different countries, as far away as Taiwan and Australia.

It became obvious during the preparation of the congress and more so at the event itself, that a platform for information in all its functionalities is desperately needed. As was to be expected, the moments of discovery of similarity in the use of ICT between the various fields were revealing. At these instances the "syndrome" of the reinvention of the wheel became apparent.

3 ICMCC, the Council

The concepts that initiated the 2004 Event became the starting points of the newly founded council, a central place where as many aspects of medical and care ICT and networking (compunetics) could come together in many different ways. Out of that concept, the following goals emerged.

3.1 Goals

The central objective of ICMCC is to create a global technology-based knowledge infrastructure that serves as:

1. a global knowledge (transfer) centre
2. a centre of expertise
3. an information dissemination platform
4. a center of excellence
5. an incubator and
6. an innovation exhibition

3.2 Global Knowledge Centre

Organizations like Healthwise in the US (www.healthwise.org) with its millions of users per year show the necessity as well as the benefit of delivering appropriate information to patients/consumers. According to its CEO, Don Kemper, "Consumers ... helped save between $7.5 and $21.5 million by avoiding unnecessary ER and doctor office visits" [2].

The availability of information works on both the B2B and the B2C level, as the structure will aim at both the professionals (caregivers) and the consumer. Professionals will be able to find relevant information (medical, technical, scientific) in a fast and efficient way. Industry (and more specifically SMEs) will have access to technical information from a central portal. Patients/consumers will be able to obtain information related to their illness or handicaps such that they will be more knowledgeable about possible treatments and treatment alternatives. The shifting paradigm of health from reparative to preventive will enhance the necessity of consumer-related information, that, when efficiently obtained, can be of great economical benefit.

In a world where the need for care is growing rapidly and where it is impossible to expect a growth in the number of caregivers, information is becoming more and more crucial. Not only because an informed patient is an economic benefit, as said before, but also because awareness amongst professionals about developments in their own and related fields can save enormous amounts of money. An example is the field of telehomecare in Europe. A growing number of projects can be found both regionally and nationally. Since most of these projects do not know of each other's existence, almost all of them follow, up to a large extent, similar protocols. Centrally available information might help to save considerable amounts of funding, because the previously mentioned reinvention of the wheel can be minimized.

The knowledge centre will be realized as a system of systems.

3.3 Centre of Expertise

ICMCC will build a global network of professionals in medicine and care. Clinicians, pharmacologists, managers, care practitioners, patients, policy makers, IT specialists, all will be represented on national and international levels within the ICMCC organization, thus providing the world with an important network structure that can be used for advisory and counseling purposes.

3.4 Dissemination Platform

Fundamental to the structure of ICMCC is the dissemination of information. There is a need for a central platform for many organizations and initiatives. Many of the largest umbrella organizations in the world lack a platform where all the various aspects of medicine and care in relation to ICT can be integrated.

Awareness will be one of the key words within the description of the ICMCC mission. Patient awareness seems an obvious goal, but also amongst professionals one can see the need. Many clinicians still see ICT (computers) as a thread to their existence and not, as it should be in our view, as a tool towards efficiency, in time as well as in costs, but also in treatment [3].

In Germany, the insurance foundation for miners (Bundesknappschaft) started a trial in 1999 in which they linked ("vernetzen"), with the help of ICT, both general practitioners and clinicians and delivered a "Gesundheitsbuch" (health book) to patients. The reason why they started this trial in the Bottrop area was because 20% of the insured caused 80% of the expenditures. In the third year (2001) the savings in costs were 7%, and the average number of days spent in hospital decreased from 12 to 8.9 [4].

In addition to its role as a dissemination platform, ICMCC will independently serve as a meeting and discussion platform for any and all parties involved in medical and care compunetics.

3.5 *Centers of Excellence*

As stated in its goals, ICMCC will help to stimulate research in a number of areas as well as bring the experts together. Across the world a limited number of highly specialized centers will be created in cooperation with industry and universities.

3.6 *Incubator*

As much as ICMCC can stimulate research, the council can also be instrumental in bringing together research and industry (especially the SMEs). Here as well we want to act as a link between the various, national incubator facilities.

3.7 *Innovation Exhibition*

ICMCC will also serve as a window to the world of ICT-related innovations in the medical and care fields in the way of an exhibition where both research and industry can jointly show there latest results.

4 The ICMCC Event 2005

ICMCC was started as a means to show the synergies in medical and care compunetics (the fact that this synergy did and does not seem obvious was the reason why). While writing this article, a discussion has been going on between some of the chairs of the ICMCC Event 2005 as to which paper/workshop should be part of which symposium.

This discussion demonstrates the effectiveness of the ICMCC concept. The proposals were delivered by the authors themselves to a specific symposium, e.g. the symposium on e-health. But looking at the various inputs it became clear that a classification was not that easy to make. Some papers deal for a large part with standardization more than with e-health, others could as well be scheduled within the symposium on information management.

Some of the symposia clearly illustrates the role of ICMCC as an international discussion platform, especially the presentations on e-health and the virtual hospitals. The latter is one of the first in western Europe on this issue. Essential for both discussions is the change in the perception of concepts that is actually taking place. What is the difference between e-health, tele-health and tele-medicine? Is there any difference? Should the concept of the virtual hospital really be called that way? Does it have any relationship with a "building"? And what will be the benefit for the patient in these concepts? To what extent will the type of patient, influence the definition of a concept? It might very well be that the outcome of the discussion on virtual hospitals might result in varying definitions depending on whether one is talking about a soldier, a rural citizen or an urban citizen, or maybe even a handicapped or elderly person.

We have been very proud that so many outstanding key-individuals in the medical and care fields have joined the ICMCC initiative. During our first meeting at the 2004 Event, there was a lively discussion on whether the Event should focus on specific subjects. The Event board had the wisdom to decide that it would be far too early to do so. They agreed with ICMCC's founder that crystallizing at this stage would deliver a massive rock that would lack all the flexibility that was at the base of the initiative. Out of that "freedom" the council was founded. This year's Event as well as the rapidly growing international recognition shows how wise that decision has been.

References

1. Bos L, Laxminarayan S, Marsh A. Medical and care compunetics 1. Amsterdam: IOS Press; 2004.
2. Kemper DW, Mettler M. Information therapy. Boise: Healthwise; 2002. p. 133
3. Kopec D, et al. Errors in medical practice: identification, classification and steps towards reduction. In: Medical and care compunetics 1. IOS Press; 2004. p. 126ff.
4. Müller H. Gewinnen durch Kooperation, Aerzte Zeitung. Accessed 13 Nov 2002.

Healthcare Compunetics

Andy Marsh, Swamy Laxminarayan and Lodewijk Bos

Abstract Changes in life expectancy, healthy life expectancy and health seeking behaviour are having an impact on the demand for care. Such changes could occur across the whole population, or for specific groups. Changes for specific groups will be particularly affected by policy initiatives, while both these and wider changes will be affected by people's levels of engagement with their health and the health service itself. Levels of education, income and media coverage of health issues are also important. These factors could also encourage an increase in people caring for themselves and their families or community. People are now expecting a patient-centred service with safe high quality treatment, comfortable accommodation services, fast access and an integrated joined-up system. The uptake of integrated Information and Communication technologies (ICT) will be crucial. Healthcare Compunetics, the combination of computing and networking customised for medical and care, will provide the common policy and framework for combined multi-disciplinary research, development, implementation and usage.

Keywords Compunetics · Electronic Patient Record · Self-care

First published in: Bos L, Laxminarayan S, Marsh A, Editors. Medical and Care Compunetics 1. IOSPress; 2004. p. 3-11. ISBN 978-1-58603-431-3.

A. Marsh (✉)
VMW Solutions Ltd, 9 Northlands Rd, Whitenap, Romsey, Hampshire S051 5RU, UK
e-mail: andy.marsh@vmwsolutions.com

S. Laxminarayan
Idaho State University, Institute of Rural Health, Pocatello, USA

L. Bos
EFSICT Foundation, Stationsstraat 38, Utrecht 3511 EG, The Netherlands

Commun Med Care Compunetics (2011) 2: 27–39
DOI: 10.1007/8754_2010_13

Published Online: 30 March 2011

1 Introduction

During the last 5 years telemedicine has utilized developing technologies and matured into a now usable service acceptable both by patients and medical staff. In essence telemedicine supports the remote application of healthcare services. Isolated medical centres can be connected to hospitals, ambulances can transmit vital sign data to awaiting emergency units, General Practitioners can be kept informed of hospitalised patients and outpatients can be monitored whilst at home. By utilizing the latest wireless technologies a new collection of "wireless telemedical" services can be developed targeting self-care and well-being applications. These new services will support not only home care services but also mobile care services for example an outpatient may go about their daily business but still have the confidence they are being continuously monitored.

Patients and public expectations of future healthcare are changing. Further enhancements to quality beyond those presently planned will be required and patients will demand provision of greater choice. Additionally, there is the changing needs of the population including demography. Over the next 20 years, the changing age structure is likely, especially for the older people, to demand more from the healthcare service.

> *"The balance of health and social care is still skewed too much towards the use of acute hospital beds. More diagnosis and treatment should take place in primary care. There is scope for more self-care"*
> Derek Wanless (Securing our Future Health: Taking a Long-Term View, April 2002)

In the future patients will be at the heart of the health service with access to better information, involved fully in decisions—not just about treatment, but also about the prevention and management of illness. The service will move beyond an 'informed consent' to an 'informed choice' approach. In this vision, patients receive consistently high quality care whenever and whoever they are. Different types of care are effectively integrated into a smooth, efficient, hassle-free service. With support from the medical institutions, people will increasingly take responsibility for their own health and well-being.

The degree to which self-care becomes more important over the next 20 years will depend on the degree to which the public engages with health care. It is therefore closely linked to some of the other trends associated with rising knowledge, such as improved public health and increased health seeking behaviour.

Self-care is one of the best examples of how partnership between the public and the health service can work. The health service can support a pro-active public in promoting self-care by, for example, helping people to empower themselves with appropriate information, skills and equipment or supporting people to take a more active role in the diagnosis and treatment of a condition followed by rehabilitation and maintenance of well-being.

A comprehensive strategy on self-care would attempt to incorporate a wide range of approaches and models of self-care, to be combined to provide safe, high quality treatment patient centred services with integrated joined-up systems with fast access.

Healthcare Compunetics, the combination of computing and networking technologies customised for healthcare, can provide the supportive underlying platform, facilities, equipment and technology to support self-care development. Healthcare compunetics is not just about home monitoring with handy, wearable devices. The most significant innovation is that, at all time, it will bring together the medical professionals with the patient and their family and carers. Healthcare compuentics will open new ways for collaboration and information sharing in health provision, something which is now barely available. Healthcare compunetics will manage the information flow and the necessary actions of all people involved in an unprecedented way for medical and care services. And all this will be achieved in a user-friendly virtual environment, within the reach of all actors, including and above all the patients, whether at work, at home or on vacation—indeed everywhere at anytime—while maintaining the privacy of all actors and the confidentiality of the medical record.

The market prospects for healthcare compunetics are very significant. The concept of healthcare provision at the point of need has expanded dramatically of the past quarter of century. Nowadays, and thanks to the advances of medicine and medical apparatus, it is common practice for long-term patients to live a normal life and be catered for by specialised staff at their home. Remote monitoring is already part of some people's daily routine—for instance cardiac patients, who may take an ECG of themselves and transit it to their doctor across a regular line. Certainly today's systems look primitive compared to what is achievable even with current technology.

Healthcare Compunetics consists of *intelligent* EPR's, *intelligent* compunetics and *intelligent* services. Presented in the next three sections, the concept of an advanced electronic patient record (EPR) is introduced in Sect. 2 and in Sect. 3 the movement of the patients data with leading edge computing and networking technologies is presented. By adopting advanced networking and computing technologies and interoperable data representations the foundations are provided for the development and implementation of advanced services as addressed in Sect. 4. For worldwide acceptance of the potential and benefits of healthcare compunetics it needs to be based on standards with well-defined interfaces. This issue is addressed further in Sect. 5.

2 The Intelligent Electronic Patient Record (*i*-EPR)

Advanced networking and communication technologies have provided the platform to sustain an Electronic distributed *hyper*-linked version of the Patient Record. Containing all the patient's medical data (collected in medical institutions and verified by medical staff) this concept can be taken one stage further to include health data that has been collected by the patient themselves, i.e., has not been verified by medical staff. This data, referred to hereafter as notes as opposed to records, can provide valuable information of historic trends and present more information for the doctor's decision, for example a regular home monitoring of blood pressure could identify a trend towards hypertension.

A personal medical data reading, such as blood pressure, collected by the patient can be regarded as a packet of information. Each of these packets can be stored in XML format as an "intelligent note" or *i*-Note for short. The note is intelligent because it can have an application associated and stored with it and it can also be viewed from different perspectives depending on the viewer's characteristics (i.e. doctor, patient, carer). Intelligent notes (*i*-Notes) are data items (blood pressure readings, temperature readings, etc.) either collected under the patients control or automatically recorded through intelligent interfaces to measuring devices such as weighting scales. *i*-Notes are not restricted to ASCII characters and may contain multimedia data such as movies and pictures. *i*-Notes, analogous to files in a traditional computing system, can be grouped together and referred to collectively as *i*-Pads. *i*-Pads are analogous to folders in a traditional computing system, however, the process of creating *i*-Pads depends on the viewer's characteristics. Different viewers may view the same *i*-Notes as different *i*-Pads. Each *i*-Pad can have an application associated with it to pre-process data. Designed originally for healthcare purposes, *i*-Notes and *i*-Pads are equally applicable in any remote monitoring environment. The concept of *i*-Notes provides the flexibility to interface to a wide variety of platforms and legacy systems.

Additionally, since a major trend nowadays is to have a personal mobile phone it makes sense also to have a limited amount of emergency information (allergies, blood type etc.) stored on a predefined area of the SIM card located in the mobile phone which in the case of an emergency could be accessed by medical staff.

The intelligent Electronic Patient Record (*i*-EPR) therefore consists of three data records linked together:

Fig. 1 *i*-SIM

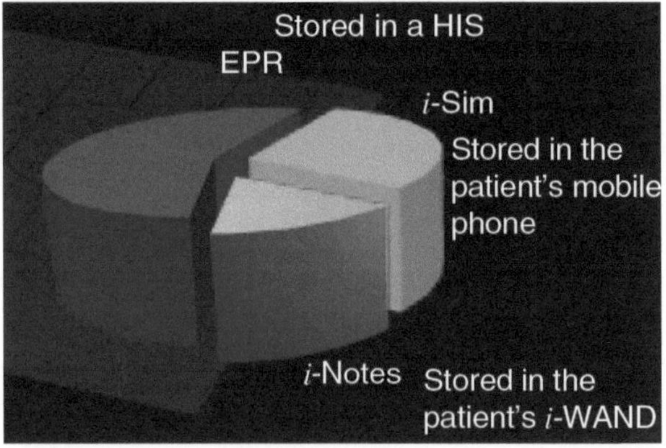

- The *i*-SIM, stored on the patient's mobile phone, which contains the patient's emergency information such as blood group, allergies, insurance details, etc. (Fig. 1).
- The traditional EPR that contains the patient's medical records that have been verified by a medical organisation.
- The *i*-Notes, that contains data items collected by the patient and third party services stored on the patient's *i*-WAND (see Sect. 3.1).

As introduced in the next section advanced communication and computing technologies can be employed to collect, transport and analyse the patient's data in a transparent unobtrusive manner.

3 Intelligent Compunetics

Modern and integrated information and communication technologies (ICT) can be used to full effect, joining up all levels of health and social care and in doing so deliver significant gains in efficiency. For example, repetitive requests for information can be avoided as health care professionals can readily access a patient's details through their EPR. As depicted in Fig. 2 and detailed further in Table 1, *i*-Compunetics combines 21st century computing and networking technologies to provide a platform to support advanced *intelligent* healthcare data collection and communication devices, such as the *i*-WAND, *i*-Port and *i*-Server.

By associating intelligence with the patient notes the respective devices can also be made to be intelligent, for example an *i*-WAND can perform analysis and diagnostics as it collects and stores the patient's data, an *i*-server can archive the patients data that has been sent by SMS, check for alarm conditions and perform more extensive data analysis and an *i*-port can transmit data directly from a measuring device, such as a blood pressure meter directly to an *i*-server.

Fig. 2 *i*-Compunetics

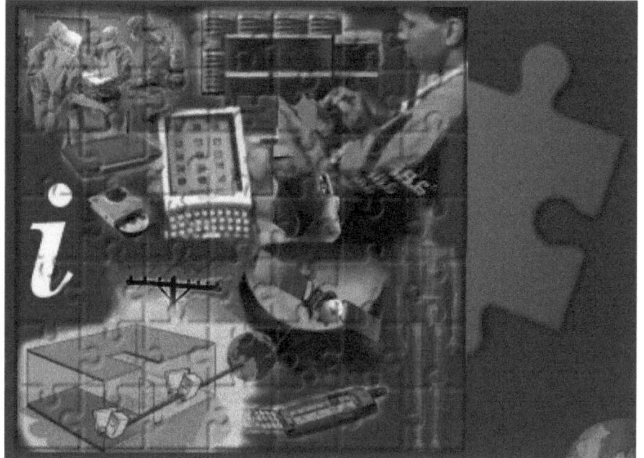

Table 1 *i*-Compunetics joining together the puzzle of ICT technologies

Wireless hospital area network (WLAN)	The interconnection of hospital information systems (HIS), picture archiving systems (PACs) and medical devices within the hospital environment. Medical staff can have wireless access to medical services and patient data whilst in the hospital environment
Wireless medical area network (WMAN)	Medical staff can access patient data outside the medical establishments (telemedicine). Fast access can be used to support movements of patient data and tele-monitoring services
Wireless home area network (Home WLAN)	Outpatients can be monitored in their home with a local network of connected monitoring devices
Wireless personal area network (PAN)	Short-range wireless communications can be used to collect personal *i*-sensor data whilst the patient is on the move
Satellite	Digital television (DTV) and GPS services can be employed to support homecare and patient on the move services
GSM/GPRS/UMTS	Mobile telecommunications technologies can support patient on the move services and remote access
Intelligent sensors	Miniature disposable wireless transmitting sensors can collect personal vital signs data.
Intelligent devices	Medical devices can perform on-the-fly analysis of patient data related to a patient's profile.
Intelligent clients	Client applications customised for the patient can be monitoring vital signs data trends.
Intelligent PDA's	PDA's can be used to provide customised viewing of patient data.
Intelligent servers	Servers can be used to perform trend analysis and data mining analysis of patient data
Intelligent mobile phones	Long-range communication of patient data can used whilst the patient is on the move.

3.1 **i-*WAND***

An Intelligent Wizards for Analysis, Note-taking and Diagnostics (*i*-WAND) is a hand-held intelligent storage and processing device with fingerprint authentication. Personal data stored on an *i*-WAND are in *i*-Note format. Each *i*-Note can have an associated Java wizard to process data on the fly and check for alarm conditions. All data saved on the *i*-WAND is automatically encrypted and hidden making it secure and unexposed. The Java wizards automatically detect which device the *i*-WAND is connected to, automatically analyse the data as it is being recorded as *i*-Notes and automatically performs diagnostics for alarm conditions (Fig. 3)

Fig. 3 *i*-WAND

3.2 i-Socket

An *i*-Socket provides intelligent access to supported medical devices. An *i*-Socket allows the *i*-WAND to be connected to a variety of data collection devices including:

- Biocompatible sensor chips in ingestible capsules
- Flat padded water resistant hypoallergenic dermal patch
- Homecare monitors

3.3 i-Server

The *i*-Server designed originally for healthcare tele-monitoring, but applicable in any remote monitoring environment, provides a stand-alone server with the capabilities to send, receive and process SMS messages. The *i*-Server provides a complete interface to specify, request, record and view the tele-monitored data. The *i*-Server can transmit alarms to support its tele-ambulatory *i*-Services and reminders to support its conformance *i*-Services. Messages are automatically stored in device folders per message type. A copy of the received messages can also be copied automatically to a removable flash drive (Fig. 4).

3.4 i-Port

The *i*-Port is an intelligent GSM modem that when receiving a data value from a connected medical device can automatically inform a server via SMS of not only the reading but also an indication of which person the reading belongs to. Homecare devices are connected to the *i*-Port directly or via a Home Area network. Designed originally for intelligent personal health-care services the *i*-Port is equally applicable in any tele-monitoring environment (Fig. 5).

Fig. 4 *i*-Server

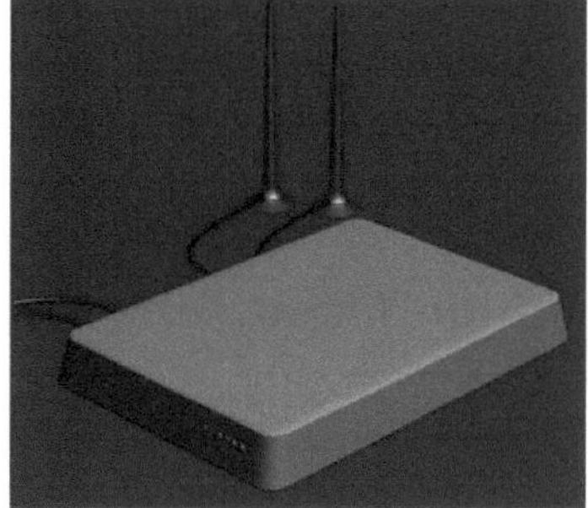

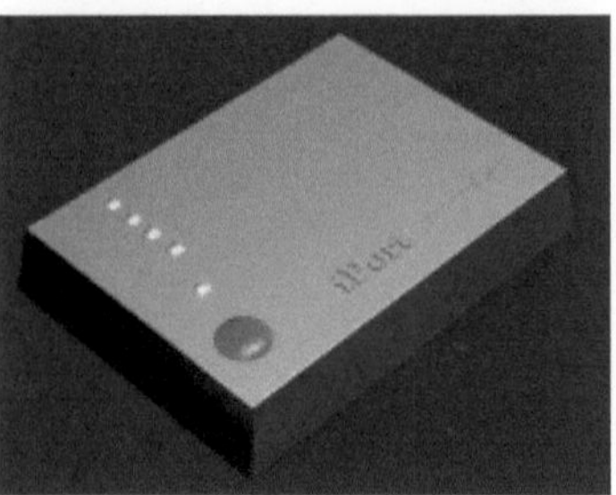

4 Intelligent Services

The *i*-Services utilising the *i*-compunetics platform can be divided into three categories:

- The Intelligent-Safety (*i*-Safety) services are a form of the newly developing mobile location services (MLS) but focus on the potential safety implications of location awareness, which include child monitoring, location advisory and third party location monitoring services.
- The Intelligent-Healthcare (*i*-Healthcare) services focus on the collection and interpretation of personal medical sensor data, which include the recording of personal sensor data (for example ECG) for future comparison, a mechanism to check the data for warning signs (for example high blood pressure) and an automated analysis of the sensor data.
- The Intelligent-Medicine (*i*-Medicine) focus on 2-way communication between diagnostic medical servers (supported by medical staff) and the users medical sensors, which support personalised care (for example informing the user, then checking that drugs have been taken with the correct dosage and at the correct time), personalised nursing (for example altering a drug prescription due to updated sensor data) and personalised doctoring (for example modifying a treatment plan).

Intelligent data and devices are only as good as the services that use them. Intelligent services can be developed that integrate a number of respective components for example a drug conformance service can utilise an *i*-server to log compliance to a drug programme and a PDA based pain assessment monitor can be reminded by an *i*-server that a reading is required. Collectively all the intelligent devices, services and data elements can be combined

4.1 Intelligent Drug Conformance Monitoring

Designed originally to support drug conformance monitoring, but equally applicable in any compliance monitoring environment, the intelligent Message Centre

Fig. 6 *i*-MC

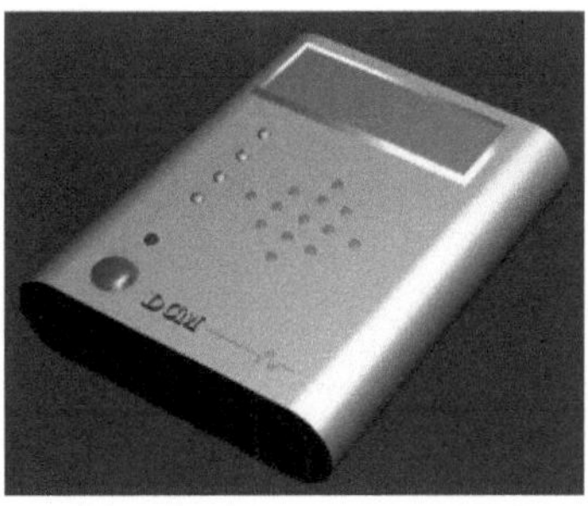

(*i*-MC) accepts "Reminder" SMS messages from a service center which then initiates a sequence of audio alarms and LED lights to inform the user that a message has arrived and requires acknowledgement. When the acknowledgement button is pressed the content of the reminder message is displayed on the LCD screen and using the latest Text-To-Speech technology, the user can listen to the message content being read.

The Drug Conformance Monitoring (*i*-DCM) service combines automated reminding and compliance logging. The rationale being that if a user acknowledges a reminder message then compliance can be assumed. If Non-compliance is indicated the server will automatically initiate an alternative action that may include an alternative means of reminding (friends, family members etc.) or even call-centre intervention (Fig. 6).

4.2 Intelligent PDA Applications and Services

The *i*-PDA, developed in Java, is a complete SMS server running on a PDA. The *i*-PDA toolkit can be used to create applications such as *i*-PAM to support remote tele-monitoring applications and *i*-Profiler to support remote access and control of an *i*-Server. The Intelligent Pain Assessment Monitor (*i*-PAM) is triggered by receiving an SMS reminder, then the user, normally the child carer, indicates by using the touch screen, which picture best describes the child's pain level. An SMS is then sent to the server and the patient's record is updated (Fig. 7).

4.3 MERLIN

Medical electronic records logistics interface to notes (MERLIN) uses the latest developments in healthcare compunetics to address social care by combining wireless communication technologies, intelligent sensors, intelligent products and intelligent services for improved self-care and non-intrusive tele-monitoring. The patients taking a more active role in their healthcare management for example can use an *i*-SIM card in their smart mobile phone to store an emergency version of

Fig. 7 *i*-PAM

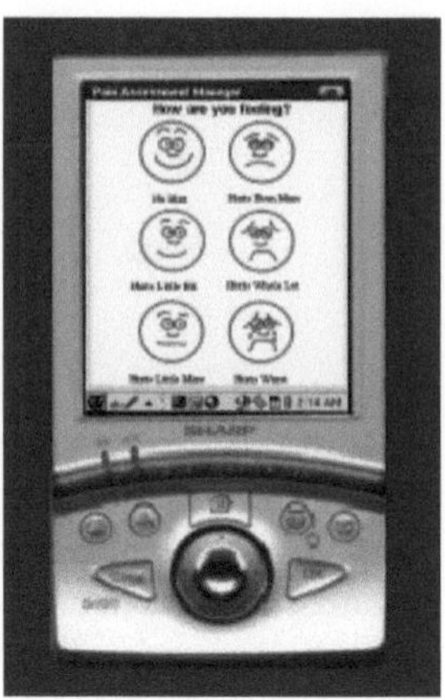

their EHCR and also an *i*-WAND (an intelligent data logging device) to collect their physiological data over time as a series of intelligent notes (*i*-Notes) thus providing a movie of their health rather than snapshots. In XML format these *i*-Notes can be used to complement the distributed patient EHCR.

MERLIN addresses not only outpatients but also includes for example patients with long-term illness, disabilities and the "Well-worried" (healthy and health conscious) with three categories of intelligent services namely, *i*-Safety, *i*-Healthcare and *i*-Medicine. The *i*-Safety services are a form of the newly developing mobile location services (MLS) but focus on the potential safety implications of location awareness such as child monitoring and guiding a blind person. The *i*-Healthcare services focus on ambulatory services via tele-monitoring both for at-home and on-the-move users, and *i*-Medicine focuses on automated 2-way digital communication between patients and carers whereby treatment plans can be analysed and adjusted remotely (Fig. 8).

Fig. 8 *i*-MERLIN

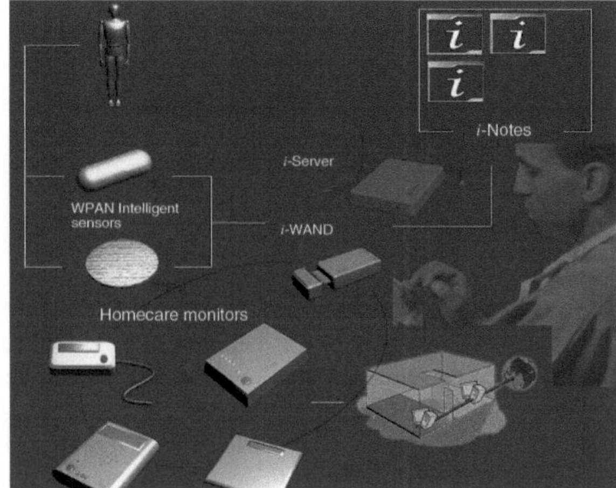

5 Conclusions

Healthcare Compunetics will play an important role in personal healthcare management. Subsequently, the development of supportive products and services will also create a new niche market economy for universities and companies, especially Small and Medium Enterprises (SME's), to develop a range of collaborative technologies.

The necessity of having standards is well understood and highly appreciated. A number of standardisation organisations and committees have been quite active in the development of standards that relate to healthcare informatics:

- American National Standards Institute (ANSI)
- CEN (Comité Europeén de Normalisation) Technical Committee 251
- ISO Technical Committee 215
- American Society for Testing and Materials Committee E31 (ASTM E31)
- Healthcare Informatics Standards Board (HISB)
- Computer-Based Patient Record Institute (CPRI.).

For the successful development of intelligent healthcare services there also needs to be an agreement on what types of categories of services will become available, how different complementary industries can work together to develop these services and how the developed services are certified.

5.1 A US Perspective

The Health Management Organizations (HMO's), that have supporting legislation, have driven the adoption of telemedicine within the US. Internet based applications are being used to improve access to care and the quality of care, reducing the costs of care and the sense of professional isolation for some healthcare practitioners. In this environment the introduction of wireless telemedicine should be introduced via the HMO's expanding their range of services and therefore compatible with the presently installed systems and envisaged wireless LAN systems. In summary, it is envisaged that the introduction of intelligent healthcare services in the US will again be driven from the HMO's therefore it is essential that some form of standardization and conformance be undertaken in conjunction with existing telemedical services.

5.2 A European Perspective

Within Europe telemedicine has not been driven so much by HMO's but more by isolated medical institutions and regional trails. The legislation aspect of telemedicine also within Europe is more complex than in the US especially when National boundaries have to be crossed. However, the telecommunications markets in Europe are more focused and standardized than their US counterparts and it is this that is envisaged to be the driving force behind the introduction of intelligent healthcare services in Europe. In summary, it is essential that some form of standardization and conformance be undertaken in conjunction with developing telecommunication infrastructures and services.

5.3 The Foundations for a Healthcare Compunetics Special Interest Group

By combining the two perspectives above it is clear that for intelligent healthcare services to be generally available and accepted worldwide then there needs to a standardization and/or conformance certification group. Similarly to the bluetooth Special Interest Group it is therefore proposed that a Healthcare Compunetics Special Interest Group (SIG) be established with representatives from both HMO's and telecommunications domains both in Europe and the US. Additionally there also needs to be representatives from a number of supporting industrials including platform developers, compunetics (computing and networking) suppliers, security advisors, medical data sensor developers and service developers.

The objective of a *Healthcare Compunetics SIG* could be to combine 20 areas of expertise:

1.Wireless medical devices (ERM TG30)	2.Compression (JPEG 2000, Wavelet)
3.Mobile terminals (PDA, Smart phone)	4.Archival (HIS, Data warehousing)
5.Operating systems (Linux, Palm OS)	6.Knowledge discovery (personalized alarms)
7.Data storage (M-EHCR)	8.Healthcare providers (Doctors, Nurses)
9.Data encoding (XML, WML)	10.Personal healthcare management providers
11.Programming environments (JAVA)	12.Standardization (R&TTE)
13.Visualization (MPEG, VRML)	14.Conformance (FDA, EU CE Marking)
15.Transmission (GPRS, EDGE, UMTS)	16.Legislation (National, EU polices)
17.Collaboration (SMS, WAP, HTTP)	18.Service providers (HMO's)
19.Privacy and security (TLS, SSL, PKCS)	20.User groups (Elderly, Outpatients)

The Healthcare Compunetics SIG will therefore tackle such issues as device availability, possibilities for health with 3G networking, the services that will be required (by health professionals, ambulatory, patients and citizens), the applications that will be developed, the costs (private and public) and compliance with technical issues, legislation and regulatory frameworks.

Bibliography

1. Marsh A, Grandinetti L, Kauranne T. Advanced Infrastructures for future healthcare. Amsterdam: IOS press; 2000.
2. http://www.vmwsolutions.com

3. Marsh A, May M, Saarelainen M. Pharos: Coupling GSM & GPS-TALK technologies to provide orientation, navigation and location-based services for the blind. IEEE ITAB-ITIS 2000, Washington Nov, 2000.
4. Strouse K. Strategies for success in the new telecommunications marketplace. USA: Artech House, 2001.
5. PriceWaterhouseCoopers. Technology Forecast 2001–2003, 2001
6. Dornan A. The essential guide to wireless communications applications. USA: Prentice Hall, 2001.
7. Jupiter Research, Mobile Commerce Report. 2001.
8. Scheinderman R. It's time to get smart. Portable design, March 2001, 22–24.
9. ARC Group, 3G Industry Survey. 2001.

Foreword: Clinical Knowledge Management: Opportunities and Challenges

Swamy Laxminarayan

I am honored to be invited by the editor, Dr Raj Bali, to write the foreword for this book. In today's information technology world, we are facing daunting challenges in realizing an all aspiring and an all encompassing paradigm of 'data-information-knowledge-intelligence-wisdom'. In the early nineties, under the aegis of the United States National Information Infrastructure, the internet facilitated the creation of an "information-for-all" environment. Despite the unstructured nature of its existence, the internet has seen an unprecedented global growth in its role as a promoter of information solutions to the citizens of the world. In contrast to the developments we witnessed in the past decade, the features of the next generation internet have shifted emphasis from the 'information-for-all' environment to a "knowledge-for all' paradigm. Some have even called it *the Internet 3*. Healthcare is undoubtedly one of the major areas in which we are beginning to see revolutionary changes that are attributable to the emergence of the knowledge engineering concepts. *Bali* and his eminent authors have done great justice to the book's contents, by pooling together many different dimensions of knowledge management into this book.

'Knowledge' is the key phraseology that has become the guiding *mantra* of future systems. As aptly stated by the National Library of Medicine's report on the next generation of their program on the Integrated Advanced Information Management Systems (IAIMS), "if the challenges of the 20th century IAIMS was tying together all of the heterogeneous systems that an organization owned, the principle challenge of the next generation of IAIMS efforts is effective integration of information, data, and knowledge residing in systems owned and

---

Originally published in Rajeev K. Bali (2005) Clinical Knowledge Management: Opportunities and Challenges, Idea Group Publishing

Commun Med Care Compunetics (2011) 2: 41–43
DOI: 10.1007/978-3-642-19554-9
© Springer-Verlag Berlin Heidelberg 2011

operated by other organizations". There is no doubt that, in recent times, we are beginning to see that knowledge revolution. Advances in the field of medical informatics are a clear testimony of newer technology developments facilitating the storage, retrieval, sharing, and optimal use of biomedical information, data, and knowledge for problem solving. These are reflected in the design and implementation of comprehensive knowledge-based networks of interoperable health record systems. They provide information and knowledge for making sound decisions about health, when and where needed.

This book delves into the technologies of knowledge management beginning from the concepts of knowledge creation and extending to the abstraction and discovery tools, as well as integration, knowledge sharing and structural influences that need to be considered for successful decision making and global coordination.

There are three major and somewhat overlapping areas of knowledge engineering applications which have dominated the healthcare sector; education, patient care and research. Knowledge stimulates creation of new knowledge and the management and dissemination of such new knowledge is the key to the building of modern educational infrastructure in medicine and healthcare. Whether it is the utility of the electronic cadaver in anatomy education, or the capturing of evidence-based medical content, or the design of a rule-based expert system in disease diagnosis, technology developments have stayed focused on creating the knowledge discovery tools, with insights mainly borrowed from the Artificial Intelligence methodologies. These include machine learning, case-based reasoning, genetic algorithms, neural nets, intelligent agents, and stochastic models of natural language understanding, as well as the emerging computation and artificial life. The central dogma in healthcare research is to ensure the patient to be the principle focus, from diagnosis and early intervention to treatment and care. Especially with the advent of the internet, clinical knowledge management is a topic of paramount importance. As *Bali et al* have pointed out in the opening chapter of this book, "future healthcare institutions will face the challenge of transforming large amounts of medical data into clinically-relevant information for diagnosis, to make recognition of it by deriving knowledge and to effectively transfer the knowledge acquired to the caregiver as and when required".

Creation of new knowledge from existing knowledge is what makes the field grow. *Bali* and his authors present in the book a number of discussions of the available technologies to stimulate the future expansion. Knowledge repositories are increasingly getting larger in size and complex in structure, as seen for example, in the hospital information systems. Such massive data explosions require efficient knowledge management strategies, including the critical need to develop knowledge retrieval and data mining tools. The latter mostly consist of appropriate software-based techniques to find difficult-to-see patterns in large groups of data. The effective analysis and interpretation of such large amounts of data collected are being enhanced by applying machine vision techniques while at the same time we are looking at machine learning mechanisms to provide self-learning instructions between processes. These are all some of the modern day innovations that are providing the capabilities to extract new knowledge from the existing knowledge. Healthcare is benefitting immensely from these applications, making it possible for healthcare professionals to access medical expert knowledge where and when needed.

Medical knowledge stems from scores of multiple sources. The design principles for the management of knowledge sharing and its global impact are a complex mix of issues characterized by varying cultural, legal, regulatory, and sociological determinants. What is especially important is to improve the overall health of the population by improving the quality of healthcare services, as well as by controlling the cost-effectiveness of medical examinations and treatment (*Golemati et al*). Technology's answer to this lies in the vast emergence of clinical decision support systems in which knowledge management strategies are vital to the overall design. I am very pleased that the authors have done an excellent job by taking a succinct view of what the issues are and the priorities of what needs to be addressed in this 'fast lane' knowledge world at large and the literature resource in particular. My congratulations to Editor *Bali* and all his team.

Prof Swamy Laxminarayan, fellow AIMBE
Chief of Biomedical Information Engineering
Idaho State University, USA
s.n.laxminarayan@ieee.org

Swamy Laxminarayan: Curriculum Vitae and Career Highlights

Following you will find an overview of Swamy's professional life as well as his written heritage.

Swamy's brother Dr. Rajaram Lakshminarayan gave us the CV he used when applying for his job in Idaho. Prof. Neil Piland from the Idaho State University provided us with an overview of Swamy's achievements in his last 2 years. I would like to express the editors' gratitude to both.

The two documents have been merged and the result is split into two sections, curriculum vitae and career highlights, and the bibliography. We have tried to be as complete as possible. (LB)

Executive Summary

Strong combination of cross-functional academic, industrial, clinical, pharmaceutical and senior managerial and executive experiences in engineering, computing and information technology applications to medicine, biology and health care

Areas of Training

Computer Science, EE and Medical Electronics, Measurement and Instrumentation, Digital Signals Processing, Control Systems, Cybernetics, Anatomy and Physiology, Mathematics and Statistics, Physics, Chemistry, Ergonomics, Aerodynamics, Flight Mechanics, Theory of Structures and Acoustics

Commun Med Care Compunetics (2011) 2: 45–105

DOI: 10.1007/978-3-642-19554-9

Technical Discipline Profile

Information Technology, Biomedical Engineering, Biomedical Sciences, Biomedical Information Engineering, Computational Biology and Bioinformatics, High Performance Computing and Communications, Health Care Networking, Information Infrastructure Architecture, Higher Education and Research, and Industrial R&D

R&D Expertise

Information Systems and Technology, Medical Informatics, Biomedical Signals and Image Processing and Visualization, Telemedicine and Telehealth, Molecular Modeling and Dynamics, Voice and Video Over IP, Microprocessor Based Medical Instrumentation, Medical Devices, High Performance Computing and Communications, Expert Systems and Knowledge Engineering, Mathematical Modeling

Biomedical and Clinical Research Application Areas

Cardiology, Radiology, Physiology, Intensive Care Patient Monitoring, Infant Home Monitoring, Sudden Infant Death Syndrome and Respiratory Distress Syndrome Studies, Biomechanics and Rehabilitation Engineering, Orthopaedics, Neurophysiology, Pharmacology, Computer-Assisted Drug Design, Circadian Physiology and Sleep Research, AIDS Research, Hansen's Disease, Telemedicine, Teleradiology, Genetic and Protein Engineering, and Information Technology Applications in Health Care and Pharmaceutical Industry (Bioinformatics, Drug Discovery, Clinical Trials, Electronic Submission of New Drug Applications and others).

Teaching Experience

Significant teaching experience at undergraduate, graduate and post-graduate levels in Computer Science, Information Systems, Internetworking, IP Communications Design, Biomedical Signals Processing, Physics, Mathematics, Medical Informatics, Image Processing, Artificial Intelligence and Expert Systems, Computational Biology including Bioinformatics, Molecular Modeling and Dynamics, Biomedical Computing, Medical Imaging and PACS, and Information Technology Applications in Health Care and Pharmaceutical areas.

Personal Data

Married, wife (Marijke), son (Vinod) and daughter (Malini), U.S. Citizen

Academic Appointments

- Chief, Biomedical Information Engineering, Idaho State University, Pocatello, ID, USA
- Chief Information Officer, National Louis University, Chicago, Illinois, USA
- Adj Professor of Biomedical Engineering, NJ Institute of Technology, Newark, NJ, USA
- Clinical Associate Professor of Medical Informatics, University of Medicine, NJ, USA
- Director, Computational Biology, University of Medicine and Dentistry of NJ, USA
- Senior Research Fellow, Albert Einstein College of Medicine, Bronx, New York, USA
- Visiting Professor of Biomedical Information Technology, Univ of BRNO, Czech Rep
- Honorary Professor, Engineering and Health Sciences, Tshingua University, China
- Assistant Professor, Thorax Center, Erasmus University, Rotterdam, The Netherlands
- Principal Investigator, Physiology Lab, Free University, Amsterdam, The Netherlands
- Senior Research Fellow, University of Southampton, Southampton, UK

Hospital Appointments

- Vice Chair, Medical Imaging and Visualization Group, University Hospital, Newark, NJ
- Director. Computer Center, Montefiore Hospital and Medical Center, Bronx, New York
- Research Physicist, Christian Medical College Hospital, Vellore, Madras State, India

Industry and Executive Appointments

- Executive Director, Collegis, Inc, National Louis University, Illinios, USA
- Director, Bay Networks Educational Center, Princeton, NJ, US

- Head of Health Care & Pharmaceutical Info Services, NextGen Internet USA, Princeton
- Director, VocalTec Corporate University, VocalTech Communication, Fort Lee, NJ, USA
- Director, Division of Educational Services, NextGen Internet USA, Princeton, NJ
- Aerodynamicist, Hamburg Aircraft Company, Hamburg, West Germany
- Flight Test Engineer, Hamburg Aircraft Company, Hamburg, West Germany

Professional Appointments (Sample List)

- *Founding Editor-in-Chief, IEEE Transactions on Information Technology in Biomedicine*, A Publication of the Institute of Electrical and Electronics Engineers, Inc
- *Editor Emeritus, IEEE Transactions on Information Technology in Biomedicine*, A Publication of the Institute of Electrical and Electronics Engineers, Inc
- *Board of Eminent Editors*, Journal of Applied Biomedicine, International Academy of Biomedical Sciences
- *Associate Editor*, IEEE Engineering in Medicine and Biology Magazine, A Publication of the IEEE Engineering in Medicine and Biology Society
- *Vice President* and Member At Large, IEEE Engineering in Medicine and Biology Society, IEEE, USA
- *Publications Services and Products Board*, Institute of Electrical and Electronic Engineers, Inc
- *United States Delegate* to various International Societies including the International Federation for Medical and Biological Engineering, the European Society for Engineering in Medicine, International Measurements Confederation and the International Federation for Automatic Control

Executive Management Experience

Executive Management of Large Computer and Information Technology Centers, Director of Multimillion Dollar Technology Projects, Long Range, Strategic and Institutional Planning, International Marketing and Sales, Budgeting and Project Management, Staff and Faculty Recruitment and Supervision, Inter-Institutional Committees, Excellent Oral and Communications Skills, International Liaison

Computer Hardware/OS/IT Proficiency

Zuse, ICT series, CDC6600, PDP11 series, HP1000series, HP9000 series, Silicon Graphics Challenge, Indy and Indigo platforms, Cyber205, CRAY-YMP, Compaq, Fortran, C, Assembler, Pascal, RTE, Unix, MS-DOS and Windows NT, LAN/WAN, TCP/IP, Client-Server paradigms, Intranets/Extranets, HTML and JAVA, Voice and Video Over IP Gateways and Gatekeepers, Routers and Switches, Virtual Private Networks, IP Network Planning

Software Development (Sample Projects)

Design, development and supervision of software systems involving hundreds of thousands of lines of code, during my professional career; Typically the applications have involved real time acquisition and processing of analog electrophysiological data, medical imaging and visualization projects, expert systems, DNA sequence analysis, critical care monitoring and microprocessor-based instrumentation software, gel electrophoresis image processing, complex mathematical and stochastic modeling of physiological systems, code optimization for supercomputer applications, databases and clinical information software, digital simulators in flight mechanics applications, aerodynamics, and a number of large scale engineering and biomedical systems modeling .

Honors and Awards

1985

- Outstanding Accomplishment Award *for internationalization of the Society's Annual International Conference,* IEEE Engineering in Medicine and Biology Society, Chicago

1986

- Presidents Award for Outstanding Academic Performance and Extraordinary Services to the University in Education and Research, University of Medicine and Dentistry of New Jersey, Newark, New Jersey

1987

- Outstanding Achievement Award *for Most Innovative International Technical Program Development*, IEEE Engineering in Medicine and Biology Society, Boston

1988

- Invited Guest Editor, *Biomedical Supercomputing*, IEEE Engineering in Medicine and Biology Magazine, New York

1989

- Elected Honorary Life Member of the Biomedical Engineering Society of India for *leading contributions to the development of biomedical engineering activities in India*, Bombay, India

1991

- Elected *United States Delegate* to the Administrative Council of the International Measurements Confederation, TC 13, Budapest
- Elected by EMBS AdCom as the nominee for the Richard Emberson Award for *outstanding efforts in the internationalization of the Society's technical activities*, Institute of and Electronics Engineers, Inc (IEEE), Piscataway, NJ
- Elected as a member of the *Distinguished Lecturers Pool*, High Performance Computing and Biomedical Information Technology, IEEE Engineering in Medicine and Biology Society
- Invited Guest Editor, Issue on *Computers in Medicine*, IEEE Engineering in Medicine and Biology Magazine, New York
- Awarded Honorary Life membership of the Romanian Society of Biomedical Engineering and Clinical Computing for *international leadership leading to the recognition of the Romanian Society*

1992

- Distinguished Service Award for initiating, developing and expanding the international focus of the biomedical engineering technical activities, Institute of Electrical and Electronic Engineers, Inc, Paris, France

- Elected as a member of the *Public Policy Commission* of the American Institute of Medical and Biological Engineering

1993

- *United States delegate* to the Scientific Advisory Board of the European Society for Engineering in Medicine, Stuttgart, Germany
- Career Achievement Recognition for pioneering contributions to biomedical engineering and information technology and international bioengineering leadership, IIT/AIIMS/IEEE EMBS, Delhi, India

1994

- *United States delegate* to the General Assembly of the International Federation for Medical and Biological Engineering, Rio de Janerio, Brazil
- *Elected as a Member of the Board of Eminent Editors,* Academy of Biomedical Scientists, Journal of Basic and Applied Biomedicine, New Delhi, India
- Elected as a Senior Member of the Institute of Electrical and Electronics Engineers, Inc (IEEE) for *significant contributions in the field of Electrotechnology*
- Elected *Vice President* of the IEEE Engineering in Medicine and Biology Society, IEEE
- Recipient of the Purkynje Award for pioneering contributions in the field of advanced computer applications to cardiovascular, neuro and pulmonary physiology and for international leadership in information technology in medicine, Prague, Czech Republic
- Elected as a Fellow of the Czech Academy of Medical Societies for outstanding scientific contributions in biomedical information technology
- Inducted as a Fellow of the American Institute of Medical and Biological Engineering for *outstanding contributions to the advancement of computational technologies in biomedical education and research,* National Academy of Sciences, Washington, DC

1995

- Member of the *Nominations Committee,* American Institute of Medical and Biological Engineering, Washington, DC
- Advisory Board, Handbook on Biomedical Engineering, CRC Press

1996

- Appointed *Founding Editor-in-Chief* of the IEEE Transactions on Information Technology in Biomedicine, Institute of Electrical and Electronics Engineers, Inc, Piscataway, NJ

1997

- *Board of Trustee's Award*, Industry Excellence, IT in Medicine, NGI, Princeton

2000

- *IEEE Third Millennium Award* for "outstanding contributions to the discipline of biomedical engineering and information technology", IEEE, Piscataway, NJ, USA
- Editorial Board, IEEE Press Series on Biomedical Engineering

2001

- *Editor Emeritus*, IEEE Transactions on Information Technology in Biomedicine, IEEE EMBS

2002

- EMBS Distinguished Service Award for "outstanding and extraordinary contributions and services in launching the IEEE Transactions on Information Technology in Biomedicine as the Founding Editor-in-Chief, and broadening the scope of activities in EMBS" Houston, Texas

2005

- *Outstanding Researcher of the Year* Award from the Idaho State University
- *IEEE Fellow*, IEEE,,for: "leadership in social and ethical implications to biomedical engineering" (posthumous)

Leadership Role in Major International Conferences (Sample List)

1985

- International Program Chair and Founding Chair of the International Committee, IEEE Engineering in Medicine and Biology Society (EMBS) Annual International Conference on Biomedical Engineering, Chicago, USA

1986

- International Program Chair, IEEE EMBS Annual International Conference on Biomedical Engineering, Fort Worth, Texas, USA

1987

Associate Program Chair, IEEE Engineering in Medicine and Biology Society

- (EMBS) Annual International Conference on Biomedical Engineering, Boston, USA
- International Program Chair, IEEE EMBS Annual International Conference on Biomedical Engineering, New Orleans, USA

1991

- International Program Chair, IEEE EMBS Annual International Conference on Biomedical Engineering, Orlando, Florida, USA

1992

- General Conference Co-Chair, IEEE EMBS Annual International Conference on Biomedical Engineering, Paris, France

1993

- International Program Chair, IEEE EMBS Annual International Conference on Biomedical Engineering, San Diego, California, USA
- General Conference Co-Chair, International Workshop on Mechatronics in Surgery, Bristol, UK

1995

- Finance Director, IEEE Neural Nets World Conference, Orlando, Florida, USA

1996

- Technical Program Chair, International Symposia on "Internet: Past, Present and Future" held in Mexico, Argentina, Chile, Brazil and USA

1997

- General Conference Co-Chair, IEEE International Conference on Information Technology Applications in Biomedicine, Prague, Czech Republic

1998

- General Conference Co-Chair, IEEE International Conference on Information Technology in Biomedicine, Washington DC, USA

1999

- General Conference Co-Chair, IEEE International Conference on Information Technology Applications in Biomedicine, Amsterdam, The Netherlands

2000

- General Conference Chair, IEEE International Conference on Information Technology in Biomedicine, Arlington, Virginia, USA

2003

- Technical Program Chair, Healthcom 2003, Santa Monica, California, USA

2004

- Scientific Chair, International Conference on Medical and Care Compunetics, The Hague, The Netherlands

2005

- Scientific Chair, ICMCC Event 2005, The Hague, The Netherlands!

Positions Held (Academia)

Chief, Biomedical Information Engineering, Idaho State University, Pocatello, Idaho, USA, 2002–2005

Position Synopsis

Responsible for multi-campus science and technology R&D initiatives in healthcare and medicine, including networking and telecommunications infrastructure design, and resource planning for state-wide telehealth and telemedicine applications. Head of Medical Informatics, Research programs in biomedical signals and image processing, medical devices and bioinformatics. Telehealth consulting for remote rural community hospitals in the State of Idaho. Research management, teaching, grant writing, publications and active healthcare provider interactions in the State.

Chief Information Officer and Executive Director, Office of Information Technology, National Louis University, Wheeling, Illinois, USA, 2001–2002

Position Synopsis

Chair, Office of Information Technology, oversee and direct the information technology and computing activities of the University at all of its 15 campuses distributed across six states in the US (Georgia, Washington DC, Missouri, Illinois, Florida, and Wisconsin) and four overseas sites in Poland, Germany, Italy and UK. Executive management, information infrastructure and global technology deployment for education, research and health care delivery management.

Clinical Associate Professor of Biomedical Informatics, University of Medicine and Dentistry, Newark, New Jersey, USA, 1991–2001

Position Synopsis

Graduate teaching and research in biomedical informatics, artificial intelligence, expert systems, medical devices, image and signals processing, genetic engineering, bioinformatics, and physiological control systems; Curriculum development of Masters and Ph.D. courses in biomedical computing, networking and biomedical informatics (Artificial intelligence, expert systems, decision support modules, bioinformatics and biomedical visualization); Graduate thesis advisory, Consultancy to the faculty and industry in decision support systems, AI applications in Medicine, bioinformatics and the University's IAIMS project

Adjunct Professor of Biomedical Engineering, New Jersey Institute of Technology, Newark, New Jersey, USA, 1985-Present

Position Synopsis

Graduate research and teaching in biomedical computing, and information technology applications in medicine and biology; Joint graduate teaching and research programs with the neighboring University of Medicine and Dentistry of New Jersey on advanced topics including *biotechnology and high performance computing;* Participate in the writing of joint research proposals for external grants from the State, Federal and Industrial agencies (NIH, NSF, Hewlett Packard etc); Supervision of Masters and Ph.D. graduate students, and participation in other academic and committee activities.

Visiting Professor, Department of Electrical Engineering, Technical University of BRNO, BRNO, Czech Republic 1997–2005

Visiting Professor, School of Engineering and Health Sciences, Tsinghua University, Beijing China, 1998–2005

Research collaborations in telemedicine and telehealth programs including high performance communication applications

Program Director for Research Computing & Information Services Technology, University of Medicine and Dentistry, Newark, New Jersey, USA, 1981–1995

Position Synopsis

Executive responsibility for building, modernizing and managing the University's multi campus state-wide computing and information technology resources, and applications to meet the educational and research missions of the University; Manage the IS staff and provide academic leadership to the comprehensive planning process of IT infrastructure development of the University including plans for merging disparate technologies and migrating to client-server paradigm; Developing and maintaining interdisciplinary R&D collaborations between the departments of Physiology, Surgery, Radiology, Anatomy, Biomedical Engineering, Molecular Biology, Medical Informatics and the University Hospital; Responsible for the design, development and teaching of biomedical computing and IT oriented biomedical graduate courses as well as courses for medical students and medical residents including active participation in the continuing education programs of the University; *Architect* of a number of networked information and instructional technology resources and Internet-based educational programs of the University; Pioneered creation of the Network Consortium between the University and the neighboring technological institutions for high tech resource sharing and developing networked educational programs. University liaison and consultant for Supercomputing resources and applications,

Development of international research collaborations with institutions in France, China, Canada, Japan and UK

Founding Head of Computational Biology Division, University of Medicine and Dentistry, Newark, New Jersey, USA, 1987–1995

Position Synopsis

Provide computing, information systems and scientific expertise in the management of four *Advanced Scientific Computing Application Sub-Disciplines* dealing with Genetic Engineering, Biomedical Signals and Image Processing, Computer Graphics and Visualization, and Molecular Modeling Technology; Major project *examples* included bioinformatics, computer-assisted drug design using molecular modeling and molecular visualization technologies, 3D visualization of anatomical structures in medical education, image processing of gel electrophoresis data, real time online monitoring and processing of electro physiological signals and large scale physiological modeling using high performance computing tools

Serve as a Senior Consultant (at Principal or Co-Principal Investigator level) on collaborative research grants to agencies such as the National Science Foundation, National Institutes of Health, and the Whitaker Foundation, in conjunction with the activities of other University departments as well as other neighboring technological institutions (the New Jersey Institute of Technology, Stevans Institute of Technology and the Rutgers University). Typical projects included the development of a Biomedical Supercomputing Educational Center, the development of a *Center without Walls* on Internet for biomedical sciences graduate education in *Computational Biology*, remote supercomputing utilization in applications such as molecular dynamics problems in drug design studies and other physiological modeling investigations.

Principal Research Investigator (Research Associate Professor), Physiology Laboratory, Free University, Amsterdam, The Netherlands, 1971–1978

Position Synopsis

Developing novel digital signals processing and stochastic modeling methodologies and modern control theory applications for analyzing and simulating physiological systems. Particular research focus emphasized cardiovascular dynamics, heart valve design, infant respiratory mechanics and the analysis of ballistocardiographic signals leading to the development of medical devices and clinical diagnostic aids and understanding of disease processes (awarded the *Purkynje Gold Medal*, one of Europe's highest recognitions, in 1994 for part of the contributions that came from these research activities)

Provide scientific and technical expertise and leadership in the development and management of the department's Computing Center; including responsibilities for directing the system design, applications programming, system integration and implementation activities; Grant writing (FUNGO), supervision and training of technical staff, serving as seminar faculty, and project leadership; Participate in curriculum development for the computational physiology courses

Senior Research Investigator (Research Assistant Professor), Thorax Center, Erasmus University, Rotterdam, The Netherlands, 1970–1971

Position Synopsis

Provide research and technical leadership in the development of computerized intensive care patient monitoring technologies including the development of real-time arrhythmia detection algorithms; Serve as the Project leader for specialized projects including the design, development and implementation of mathematical models for the understanding of the electrical wave propagation in the human heart; Participate in numerous academic committee activities which included serving on the organizing committee that initiated the first *Computers in Cardiology* conference.

Research Fellow and Assistant Director, Digital Signals Processing Center, University of Southampton, United Kingdom, 1966–1970

Position Synopsis

Founding Fellow and Head of Research & Development; Responsible for the Center's research and technical activities; Center which was the first of its kind in Europe was developed as a major national resource for on-line digital signals processing using the fastest computing facilities in Europe at the time; Served as consultant and solutions architect to university-wide faculty and student community and industry on digital signals processing methodologies; Research areas focussed on design and development of novel signals processing techniques including Fast Fourier Transform algorithm, recursive and non-recursive digital filtering methods, evolutionary spectral analysis techniques and non-stationary signals processing.

NASA Research Fellow, Institute of Sound and Vibration Research, University of Southampton, United Kingdom, 1965–1966

Position Synopsis

Research project to study the structural fatigue of supersonic aircrafts (specifically the Concorde) to high intensity noise. Project involved development of a laboratory based noise source to simulate the high intensity environment for testing aircraft structures.

Positions Held (Clinical and Hospital)

Vice Chair, Medical Imaging and Visualization (MIV) Research Group, Departments of Radiology and Academic Computing Center, University of Medicine and Dentistry of New Jersey, Newark, New Jersey, USA, 1988–1995

Position Synopsis

Provide inter-institutional leadership to the MIV Research Group in creating synergy between various University and hospital departments (Anatomy, Physiology, Radiology, Medicine, Pathology and Neuroscience) and neighboring technological institutions involved in the application of imaging, visualization and information technologies; Typical projects included 3D Visualization in medical education and research, high performance workstations for medical diagnostics, image fusion studies of multiple modalities, PACS and RIS requirements for the University Hospital's Division of Radiology, and developing the framework for Teleradiology and Telemedicine protocols; Building the technical and manpower resources drawn from the neighboring technological Universities (Rutgers and the New Jersey Institute of Technology)

Curriculum design and teaching radiology residents and medical students on information technology applications in radiology. Curriculum included information infrastructures, telecommunication principles, 3D visualization in radiology and issues involved in the appropriate design of Picture Archiving and Communication Systems (PACS) and Radiology Information Systems.

Director of Scientific Computing, Department of Neurology, Montefiore Hospital & Medical Center, Bronx, New York, USA, 1978–1981

Position Synopsis

Provide scientific and technical leadership to a large technology group consisting of 20 systems analysts, senior programmers, instrumentation engineers, and medical and postgraduate research associates on major NIH funded and other projects. The areas of research focussed on the analysis of continuous long term (24, 48 and 72 h) neurophysiological data in Sudden Infant Death Syndrome Research, pulmonary functions evaluation, sleep research, circadian physiology, and respiratory monitoring. Typical projects included the design and development of microprocessor-based respiratory monitoring devices and home monitoring devices, developing and evaluating disease profiles, physiological signals processing, modeling and automatic scoring of sleep-wake processes, and biomathematical and statistical analysis of clinical data.

Administration of the Center's activities in the planning and conducting of multidisciplinary biocomputational research, supervision of technical and academic staff, review and planning of budgets, writing of research grants, system design and planning of all computational activities

Research Physicist, Christian Medical College Hospital (A Prestigious American Missionary Medical Center), Vellore, India

Position Synopsis

Responsible for the Indian Council of Medical Research sponsored Experimental Laboratory in the Orthopaedics Department of the College under the chairmanship of Prof. Paul Brand; Research emphasis on peripheral circulation studies involving gait analysis, artificial limbs design, and peripheral circulation measurements in animal protocols and leprosy patients; Developed instrumentation and measurement techniques such as pressure transducers for gait analysis, thermocouples for skin temperature measurements and digital plethysmography for peripheral blood flow measurements. Designed and performed clinical and diagnostic tests in animal and patient protocols.

Positions Held (Industry)

Director and Chair, VocalTec University, VocalTec Communications, Fort Lee, New Jersey, USA, 1998–2001

Position Synopsis

Provide technical and corporate leadership in the development, management and global deployment of high technology educational programs, products and services (USA, Israel, UK, Japan, Europe, China, India) in information technology with emphasis on internetworking, Voice and Video over IP, IP network planning, voice data and multimedia convergence; Responsibilities include curriculum development, network planning, staff recruitment, faculty supervision, consultant services to major corporations and universities.

Provide leadership in the research and development of high tech communication platforms and embedded devices for applications in medicine, biology, pharmaceutical and health care applications (Video conferencing, telehealth and e-commerce applications in health care); Develop University-State-Federal-Industry partnership initiatives

Director of Health Care and Pharmaceutical Information Services, NextGen Internet USA Inc, Princeton, New Jersey, USA, 1996–1998

Position Synopsis

Identify and develop vertical markets in information technology applications with major focus on health care and pharmaceutical industry. Responsible for the design, development and implementation of a broad suite of information technology consulting programs for pharmaceutical and health care executives (drug discovery, drug design, clinical trials, electronic submission of new drug applications, and bioinformatics) to help take advantage of the ongoing and newly evolving technologies in the areas of high performance computing and communications, the National and Global Information Infrastructures, and the Next Generation Internet Initiatives; Develop University-Industry partner programs for the utilization of the IT resources in biomedical sciences education and research

Serve as a Senior Information Technology Advisor to program directors at organizations involved in health care delivery projects such as telemedicine, teleradiology, teleconferencing etc

Global Director, Division of Educational Services, NextGen Internet, Princeton, New Jersey, USA (High Tech Company which evolved from the former NSF funded John van Neuman National Supercomputer Center), 1995–1998

Position Synopsis

Responsible for the overall international operations (USA, Mexico, Argentina, Chile, Brazil Venezuela and India) of the world-wide educational, and high technology training activities of the institution, involving Internet, Internetworking and high performance computing and communications applications in engineering, medicine, health care, education and manufacturing industries; Provide consultancy and end-user IT solutions to educational, research, medical, health care and pharmaceutical organizations (Connectivity, protocol design, LAN/WAN network configurations, Intranet and Extranet developments, virtual corporate networks, electronic commerce, telemedicine, and high performance communications applications); Provide consultancy on leading edge *Internet-based educational technologies* to four-year colleges and universities; Faculty development, strategic planning, long range forecasting, and international policy development

Director, Bay Networks Authorized Educational Center (now known by Nortel Networks), NextGen Internet, Princeton, NJ, USA, 1995–1998

Position Synopsis

Setting directions for and management of the Global Educational Center working in strategic alliance with Bay Networks; Serve as Company's Corporate Contract Manager; Work closely with Bay Networks to ensure successful implementation of the various certification courses in high tech areas including Local and Wide Area Networks, Routers, Advanced IP and other relevant Bay Networks training protocols and internetworking technologies for practicing professionals; Identify, recruit and manage instructional faculty, and coordinate their technology certification programs with the Bay Networks Company

Aerodynamicist and Flight Test Engineer, Hamburger Flugzeugbau Gmbh, Hamburg, Germany

Position Synopsis

Responsible for all the analog, digital and hybrid computational activities of the Division of Aerodynamics and Flight Mechanics; R&D focus included computational fluid mechanics, stability and control, aeroelasticity and flight simulation studies in modern aircraft design; Developed a Digital Flight Simulator first of its kind in Europe at the time for studying flight performance and control behavior, which was also later used on the design of other modern aircrafts including the Air Bus. Served as a Flight Test Engineer on board an executive jet for 18 months.

Academic Milestones

1953–1958

- Bachelors (Honors), University of Mysore, Karnataka State, India (Majored in Physics, and Mathematics) *Graduated at the age of 14 years which gave me a very early career start.*

1960–1964

- Associate Fellowship Exam of the Royal Aeronautical Society, London, (Aerodynamics, Theory of Structures, Flight Mechanics and Materials) and completed graduate training at Field Aircraft Services, Croydon, London, UK

1965–1967

- Masters, University of Southampton, UK (Digital Signals Processing, Acoustics and Vibration Engineering, Cybernetics, Instrumentation, Electronics, Physiology, and Control Theory), Thesis: High Intensity Noise Simulation for Structural Fatigue Testing of Supersonic Aircraft Structures

1967–1972

- Doctoral Thesis, Design and Development of a Real Time On-Line Signals Processing System with Applications in Physiological Systems Analysis, University of Southampton, UK,

Major External Grants (Selected Titles)

1974

Co-Principal Investigator

Information Processing of Hemodynamic Parameters, Dutch Organization for Scientific Research (ZWO-FUNGO), Vrij Universitaet, Amsterdam, The Netherlands, *$250,000*

Principal Investigator

Cardiovascular Systems and Control Studies, Dutch Organization for Scientific Research (ZWO-FUNGO), Vrij Universitaet, Amsterdam, The Netherlands, *$190,000*

1978–1980

Co-Principal Investigator

Evaluation and Follow-up of Selected Respiratory, Cardiac and Neurophysiologic Parameters in Infants, National Institutes of Health, NICHD, Montefiore Hospital and Medical Center, New York, USA, *$886,930*

1984

Co-Principal Investigator

Acquisition of High Performance Graphics Work Stations for Biomedical Research and Education, Hewlett Packard Education Grant, University of Medicine and Dentistry of New Jersey, Newark, NJ, USA, *$160,000*

1987

Co-Principal Investigator

Hewlett Packard Equipment for Biotechnology Education, Hewlett Packard
Education Grant, University of Medicine and Dentistry of New Jersey,
Piscataway, NJ, USA, *$214,588*

1989

Co-Investigator

Morphometrics of Early Cervical Cancer, New Jersey State Commission on
Cancer Research, New Jersey, USA, *$88,000*

1994

Co-Principal Investigator

High Tech Equipment Grant for Biomedical Education and Research, NJ State
Equipment Leasing Fund Program, *$10.4 million*
Co-Principal Investigator
Acquisition of High Performance SGI Workstations for Biomedical Education
and Research in Genetic and Protein Engineering, Computer-Assisted Drug
Design, and Virtual Reality Applications, NJ State Equipment Leasing Fund
Program, *$160,000*

1995–2000

Worked in Industry and Developed a number of RFPs as well as Venture Capital
proposals

1999

Co-Principal Investigator

NJTC Industry Partnership Grant, IP Communications Infrastructure
Development, *$4.8 million*

2004

A Web-Based Integrated Information Management System in Autism Research, ISU Faculty Research Council, *$5,000*
Steering Committee Member
  BRIN / INBRE (Idaho Network of Biomedical Research of Excellence) NIH grant, *$1.5 million*

2005 Sep–2008 Nov

Original Principal Investigator

Idaho Bioterrorism Awareness and Preparedness Program, Health Resources and Services Administration, Bureau of Health Professions, Grant# T01HP06420, *$3,856,568.*
Swamy was instrumental in proposing this project which was approved and funded shortly before his death. The proposal was excellent and the project itself was exceptionally successful in meeting its goals and advancing the science of emergency preparedness. Much of his vision was achieved in performance of this program. And it is bringing lasting value to the population of Idaho. The project has been dedicated, by the Idaho State University Institute of Rural Health, to his memory and to the lasting power of his vision.

Distinguished Services (Sample Titles)

- United States Coordinator, Graduate Researchers Forum, IEEE - Innovation et Technologie en Biologie et Medicine, Paris, France, 1988
- United States Delegate, "Health for All by 2000" WHO Task Force 1988–2005
- Distinguished Member of the Educational Advisory Committee for Biomedical Engineering, New York Academy of Medicine, New York, 1993
- Elected Vice President of the 9000-member Engineering in Medicine and Biology Society of the Institute of Electrical and Electronics Engineers, Inc (IEEE), Piscataway, NJ, 1994–2005
- Finance Director, Neural Nets International Conference of the IEEE Neural Nets Council, Orlando, Florida, USA, 1994 (Budget $750 k)
- National Science Foundation Study Panels, National Challenge Groups, Health Care Delivery Systems, Washington, DC, 1994–2005
- Delegate, United States Activities Board of the Institute of Electrical and Electronics Engineers, Inc (IEEE), Washington, DC, 1995–2005
- National Institutes of Health (NIH) Study Panels, 1996–2005

- International PhD Examiner, Indian Institutes of Technology, Delhi, Madras and Kanpur, Indian Institute of Science Bangalore and All India Institute of Medical Sciences, Dalhousie University, Canada, Free University, Amsterdam
- Advisory Board Member, Numerous Journals and Electronic News Letters
- National Research Council of Canada Grants Study Panels
- US Delegate to numerous national and international biomedical engineering congress and promotional boards in countries including India, China, Estonia, Croatia, Hungary, Bulgaria, Romania, Czech Republic, Germany, UK, Poland, The Netherlands, New Zealand, Taiwan and Hong Kong
- Invited Member, Technical Advisory Board, Healthcare Technology, Institute of Electrical Engineers, UK

Invited Keynote, Plenary and Featured Lectures (Selected Samples: 35 out of 79)

- *Random Process Analysis Techniques in Medicine*, International Conference on Online Computing in Medicine and Biology, Borough Polytechnic, London, UK, 1970
- *Stochastic Modeling of the Ballistocardiogram*, International Conference on Ballistocardiography and Cardiovascular Dynamics, Amsterdam, The Netherlands, 1974
- *Convergence of Haar and Walsh Transforms in the Analysis of Neurophysiological Signals*, International Congress of Physiological Sciences, Budapest, Hungary, 1980
- *Critical Overview of Apnea Monitors*, 37th Annual International Conference of the Alliance for Engineering in Medicine and Biology, Los Angeles, USA, 1981
- *Biomedical Signals Processing Methods and Applications*, Featured speaker, New York Academy of Sciences, New York, USA, 1985
- *Spectral Analysis Applications in the Characterization of Sleep-Wake Patterns*, 11th Northeast Bioengineering Conference, Worcester Polytechnic, Worcester, USA, 1985
- *Sudden Infant Death Syndrome: Analysis of Long Term Electrophysiological Analog Data*, 14th International Conference of the Indian Biomedical Engineering Society, Bombay, India, 1988
- *Genome Mapping: Future Computational Challenges*, Computational Biology Symposium, IEEE Engineering in Medicine and Biology Society, New York, 1989
- *Supercomputing Applications in Medicine,* 5th International Conference of the International Measurements Confederation, Calcutta, India, 1989

- *Multispectral Analysis of Multidimensional Medical Data*, Special Symposium on State of the Art Integrated PC Based Software Systems, Asyst Software Technologies, Newark, NJ, USA, 1989
- *Computers and Information Technology in Medical Education*, Annual Conference of the Czechoslovakian Society of Biomedical Engineering, Prague, 1990
- *Human Genome Project: Computational Challenges*, New York Academy of Sciences, New York, 1990
- *Biomedical Visualization: Needs and Applications in Medical Diagnosis*, 6th International Conference of the International Measurements Confederation on "Measurements in Clinical Medicine", Sopron, Hungary, 1990
- *Global Telemedicine using INMARSAT Satellite System*, International Conference on Telecommunications for Health Care: Telemetry, Teleradiology and Telemedicine, Calgary, Canada, 1990
- *Trends and Directions in Biomedical Computing*, Annual Conference of the Biomedical Engineering Society of China, Shanghai, China, 1991
- *Biocomputing and Information Technology in Molecular Sciences*, International Conference of the IEEE Engineering in Medicine and Biology Society, Paris, France, 1992
- *High Performance Computing Applications in the Understanding of 3D Protein Folding Mechanism*, International Conference on Biomedical Engineering in the 21st century, Taipei, Taiwan, 1992
- *The Role of Visualization in Protein Modeling*, ELECTRA International, Advanced Technologies and Competitiveness, Edison, NJ, USA, 1993
- *Impact of HPCC Initiatives on Biomedical Research and Education*, World Congress on Medical Physics and Biomedical Engineering, Rio de Janeiro, Brazil, 1994
- *Internet Technology,* Special Workshop at the 16th Annual International Conference of the IEEE Engineering in Medicine and Biology Society, Baltimore, MD, 1994
- *Biomedical Engineering and the Information Superhighways for Mass Health Care*, IEEE International Conference on Mass Health Care, New Delhi, India, 1995
- *National Information Infrastructure: The Digital Nirvana for Health and Medical Research*, International Conference on Mechatronics in Medicine and Surgery, Bristol, UK, 1995
- *High Performance Computing and Communications: How Does It Serve U.S. Biomedical Engineering Community*, International BME Conference, Taiwan, 1995
- *Information Management in Medicine*, Southern Regional International Conference on Biomedical Engineering, Dayton, Ohio, USA, 1995
- *Global Information Infrastructure in Medicine*, IEEE Engineering in Medicine and Biology Regional International Conference, Christchurch, New Zealand, 1995

- *Internet, Internetworking and National Information Infrastructure*, International Conference, Biosignal '96, Brno, Czech Republic, 1996
- *Next Generation Internet Initiatives*, Global Enterprise Services World Symposia on Past, Present and Future of Internet, Sao Paolo, Brazil, 1996
- *Telecommunications in Health Care,* Telecommunications Summit, Organized by Congressman Bob Franks, Somerset, New Jersey, 1996
- *Telemedicine: Issues and Concerns*, Mini Symposia in Telemedicine, International Conference of the IEEE Engineering in Medicine and Biology Conference, Chicago, 1997
- *Impact of Internet on the Pharmaceutical and Health Care Industries*, IEEE International Conference on Information Technology Applications in Biomedicine, Prague, Czech Republic, 1997
- *Information Technology in Biomedicine: The Next Step Beyond*, VIII International Biomedical Engineering Conference of the International Federation, Cyprus, 1998
- *From Digital Dawn to the Next Generation Internet Initiatives*, IEEE International Conference on Information Technology Applications in Biomedicine, Amsterdam, The Netherlands, 1999
- Technology Transfer in Telemedicine, International Conference on Health Emergency Telematics organized by the HECTOR Consortium and the Health Telematics Office of the European Community, March 15–17, 1999, Seville, Spain
- *Technology, Medicine and the Patient: Who Benefits the Most?*, 11th Nordic-Baltic Conference on Biomedical Engineering Regional Meeting of the IFMBE and EMBS, Estonia, 2000
- *Crossroads of Global Tele-Issues*, MEDICON 2001, IX Mediterrannean Conference on Medical and Biological Engineering and Computing, Pula Croatia, 2001
- *Barriers and Bounds in Internet Medicine,* The Eighth Australian and New Zealand Intelligent Information Systems Conference, Perth, Australia, 2001
- *Angiography and Plaque Tissue Imaging*, BEACON Annual Symposium, Hartford, Connecticut, USA, October 2002
- *Obstrusive Dimensions to Bioterrorism*, BEACON Annual Symposium, Hartford, Connecticut, USA, March 2003
- *Information Engineering in Medicine and Health Care: Hype, Reality and the Future*, International Conference on Medical Informatics and Engineering, Craiova, Romania, October 2003
- *Role of Compunetics in Health Care*, (Invited Presentation) 1st World Congress on Men's Health, UNESCO, Paris, France, April 2004
- *Health Care Informatics*, (Invited Plenary) International Congress on Medical and Care Compunetics, The Hague, The Netherlands, June 2004

Thesis Advisory Committee Services
(Selected Titles, 20 out of 69)

1983

- The pre- and post-synaptic molecular aspects of synapse formation, Ph.D. Thesis, Department of Pharmacology, Graduate School of Biomedical Sciences, University of Medicine and Dentistry of New Jersey, Newark, NJ, USA

1984

- Deconvolution techniques with applications in cardiovascular systems analysis, Masters Thesis, Biomedical Engineering Program, New Jersey Institute of Technology, Newark, USA

1985

- Design of pulse duplicator in heart valve studies, PhD Thesis, Biomedical Engineering Program, New Jersey Institute of Technology, Newark, NJ, USA

1986

- Developing data communication protocol between CAREPORT and HP1000 computer for intensive care unit, Masters Thesis, Computer Science Program, New Jersey Institute of Technology, Newark, NJ, USA

1990

- Experimental model of coronary macromolecular transport in diabetes mellitus, Ph.D. Thesis, Department of Physiology, University of Medicine and Dentistry of New Jersey, Newark, NJ, USA

1991

- Modeling and prediction of AIDS progression, Masters Thesis, Department of Biomedical Engineering, New Jersey Institute of Technology, Newark, NJ, USA
- Texture analysis in two dimensional ultrasound B scans, Ph.D. Thesis, Department of Biomedical Engineering, Indian Institute of Technology, Madras, India
- Diagnostic expert system for psychiatric disorders, Ph.D. Thesis, Department of Biomedical Engineering, Indian Institute of Technology, New Delhi, India
- Intracranial pressure and temperature sensor with telemetry, PhD Thesis, Department of Biomedical Engineering, New Jersey Institute of Technology, Newark, NJ, USA
- Mathematical simulation of hippocampal neurons, Masters Thesis, Department of Biomedical Engineering, New Jersey Institute of Technology, Newark, NJ, USA

1992

- Time domain analysis of the cardiovascular system, PhD Thesis, Department of Biomedical Engineering, New Jersey Institute of Technology, Newark, NJ, USA
- Three dimensional reconstruction and visualization of dental pulp, Masters Thesis, Department of Biomedical Engineering, New Jersey Institute of Technology, Newark, NJ, USA
- Three dimensional reconstruction in medical imaging, B.S. Dessertation, Department of Electrical Engineering, Princeton University, Princeton, USA

1993

- Optimization of flow in the collecting duct during the concentrating mode and the diluting mode in a nephron population of two different lengths with the renin angiotensin system and ADH mechanism, Masters Thesis, Biomedical Engineering, New Jersey Institute of Technology, Newark, NJ, USA

1994

- Reconstruction and enhancement of medical Computed Tomography images, Masters Thesis, Department of Biomedical Engineering, New Jersey Institute of Technology, Newark, New Jersey, USA

- Computerized automatic sleep scoring, Masters Thesis, Department of Biomedical Engineering, New Jersey Institute of Technology, Newark, New Jersey, USA

1995

- Spectral analysis of eeg responses in the characterization of drug administration, Masters Thesis, Department of Biomedical Engineering, New Jersey Institute of Technology, Newark, New Jersey, USA

2000

- Group Decision Support system in medical education, PhD Thesis, Department of Management, Rutgers University and New Jersey Institute of Technologyt, Newark, New Jersey, USA

Educational Initiatives and Teaching

Developed graduate courses leading to Masters and Doctoral programs in Biomedical Computing, Biomedical Informatics, Computational Biology and Biomedical Information Technology

Undergraduate teaching in Physics, Mathematics, Computer Science and Computer Languages, University of Southampton, UK

Graduate teaching in Digital Signals Processing, Control Systems, and Instrumentation, University of Southampton, UK

Lecture Series on Biomedical Signals Processing, Systems Engineering and Instrumentation to post-graduates, Free University of Amsterdam, Netherlands and University of Medicine and Dentistry of NJ, Newark, NJ

Computers in Medicine and Medical Informatics Courses to medical students and residents, New Jersey Medical School, UMDNJ, Newark and Piscataway, New Jersey

Doctoral Courses in Quantitative Analysis and Information Processing, Biochemistry Department, Graduate School of Biomedical Sciences, UMDNJ, Newark, New Jersey

Doctoral Courses on Mathematical Methods in Pathology, Graduate School of Biomedical Sciences, UMDNJ, Newark, New Jersey

Graduate and doctoral Courses on Biomedical Computing and Information Technology, Graduate School of Biomedical Sciences, UMDNJ and New Jersey Institute of Technology, Newark, New Jersey

Graduate Courses in Biomedical Informatics, New Jersey Institute of Technology and the School of Health Related Professions, UMDNJ, Newark, New Jersey

Special Advanced Workshops in Computational Biology including Genetic Sequence Analysis, and Molecular Modeling and Dynamics Technologies

Special Courses on Medical Imaging, Visualization, and PACS, Radiology Department, University Hospital and New Jersey Medical School

Special tailored courses to Pharmaceutical industry executives on information technology applications in Bioinformatics, Drug Discovery, Drug Design, Combinatorial Chemistry, Clinical Trials, Electronic Submission of New Drug Applications and Electronic Commerce

Participation in International Conferences (Selected Meetings)

1980

- Session Chairman on *"Biosignals Processing"*, Satellite Symposium on Computational Methods of the International Congress of Physiological Sciences, Budapest, Hungary

1985

- Session Chairman on *"Computers in Medicine"*, International Conference on Frontiers of Engineering and Computing in Medicine, Chicago, USA
- Session Chairman on *"Neonatal Measurements"*, XIV International Conference of the International Federation for Medical and Biological Engineering, Espoo, Finland
- Track Chairman of *"Perinatal Monitoring Track"*, 8th Annual International Conference of the IEEE Engineering in Medicine and Biology Society, Fort Worth, Texas, USA
- Track Chairman of *"Clinical Research Models Track"*, 8th Annual International Conference of the IEEE Engineering in Medicine and Biology Society, Fort Worth, Texas,

1986

- Track Chairman of *"Biotelemetry Track"*, 8th Annual International Conference of the IEEE Engineering in Medicine and Biology Society, Fort Worth, Texas, USA
- Panel Moderator on *"Health Care Technology"*, 8th Annual International Conference of the IEEE Engineering in Medicine and Biology Society, Fort Worth, Texas, USA

1987

- Track Chairman of *"Biological Applications Track"*, 9th Annual International Conference of the IEEE Engineering in Medicine and Biology Society, Boston, USA
- Panel Moderator on *"Information Technology in Third World Countries"*, 9th Annual International Conference of the IEEE Engineering in Medicine and Biology Society, Boston, USA
- Track Chairman of *"Physiological Modeling and Simulation Track"*, 9th Annual International Conference of the IEEE Engineering in Medicine and Biology Society, Boston, USA
- Session Chairman on *"Analysis of Biological Fluids"*, International Conference on Electroanalytical Techniques and Instrumentation, Mysore, India

1988

- Tutorial Coordinator on *"Biomedical Supercomputing"*, International Conference of the International Federation for Medical and Biological Engineering and Computing, San Antonia, Texas, USA
- *Moderator*, Graduate Researchers Forum on Biomedical Engineering, Innovation et Technologie en Biologie et Medicine, Paris, France
- Session Chairman on *"Networking and Communications"*, 10th Annual International Conference of the IEEE Engineering in Medicine and Biology Society, New Orleans, USA
- Track Chairman of *"Biomedical Education Track"*, 10th annual International Conference of the IEEE Engineering in Medicine and Biology Society, New Orleans, USA
- Track Chairman of *"Cardio-Pulmonary Systems Track"*, 10th Annual International Conference of the IEEE Engineering in Medicine and Biology Society, New Orleans, USA

1989

- Moderator on *"Genome Mapping"*, Computational Biology Symposium, IEEE Engineering in Medicine and Biology Society, New York, USA
- Session Chairman on *"Microprocessor Applications in Medicine"*, 11th Annual International Conference of the IEEE Engineering in Medicine and Biology Society, Seattle, WA, USA
- Session Chairman, *Graduate Students Contest*, 11th Annual International Conference of the IEEE Engineering in Medicine and Biology Society, Seattle, WA, USA

1990

- Track Chairman of *"Biophysical Measurements Track"*, 12th annual International Conference of the IEEE Engineering in Medicine and Biology Society, Philadelphia, PA, USA
- Track Chairman of *"Bioengineering Education Track"*, 12th Annual International Conference of the IEEE Engineering in Medicine and Biology Society, Philadelphia, PA, USA
- Session Chairman on *"Biomedical Computing Technology"*, 12th Annual International Conference of the IEEE Engineering in Medicine and Biology Society, Philadelphia, PA, USA
- Session Chairman, *International Graduate Students Contest*, 12th Annual International Conference of the IEEE Engineering in Medicine and Biology Society, Philadelphia, PA, USA

1991

- Track Chairman of *"Bioengineering Education Track"*, 13th Annual International Conference of the IEEE Engineering in Medicine and Biology Society, Orlando, Florida, USA
- Track Chairman of *"Biophysical Measurements Track"*, 13th Annual International Conference of the IEEE Engineering in Medicine and Biology Society, Orlando, Florida, USA
- Session Chairman on *"Workstation and MicroprocessorTechnology"*, 13th Annual International Conference of the IEEE Engineering in Medicine and Biology Society, Orlando, Florida, USA

1992

- Session Chairman on *"Biomechanics"*, Second International Conference on Biomedical Engineering in the 21st Century, Teipai, Taiwan

1993

- Session Chairman on *"Advanced Technologies"*, ELECTRA '93, Raritan, New Jersey, USA
- Track Chairman of *"Medical Informatics Track"*, 15th Annual International Conference of the IEEE Engineering in Medicine and Biology Society, San Diego, USA

1994

- Panel Chair of *"High Performance Computing in Medicine"*, World Conference of the International Federation of Medical and Biological Engineering, Rio, Brazil
- Workshop Coordinator of *"Internet Technology"*, 16th Annual International Conference of the IEEE Engineering in Medicine and Biology Society, Baltimore, MD, USA

1995

- Session Chair on *"Neural Networks Applications"*, First IEEE Regional International Conference of the IEEE Engineering in Medicine and Biology Society, New Delhi, India
- Session Chairman on *"Computer Sciences in Andrology and Sexual Rehabilitation"*, First International Conference on Andrology and Sexual Rehabilitation, Paris, France

1996

- Track Chair of *"Medical Informatics Track"*, 17th Annual International Conference of the IEEE Engineering in Medicine and Biology Society, Amsterdam, The Netherlands
- Session Chairman on *"Medical Imaging"*, 17th Annual International Conference of the IEEE Engineering in Medicine and Biology Society, Amsterdam, The Netherlands

1997

- Theme Chair of *"Biomedical Information Technology Track"*, 18th Annual International Conference of the IEEE Engineering in Medicine and Biology Society, Chicago

1998

- Theme Chair of "Biomedical Information Technology Track", 19th Annual International Conference of the IEEE Engineering in Medicine and Biology Society, Hong Kong

1999

- Theme Chair of *"Biomedical Information Technology Track"*, 20th Annual International Conference of the IEEE Engineering in Medicine and Biology Society, Atlanta, Georgia

2000

- Track Chair *"Medical Informatics and Biomedical Information Technology"*, World Congress on Biomedical Engineering and Medical Physics, Chicago
- Track Chair *"Biomedical Information Engineering"* IEEE Pan Pacific Conference, Hong Kong

2001

- Track Co-Chair, Information Engineering, Annual International Conference of the IEEE Engineering in Medicine and Biology Society, Istanbul, Turkey

IEEE Offices Held

- 1985–1986 International Program Chair, IEEE EMBS International Conferences

- 1986–1997 Technical Program Committee, IEEE EMBS International Conferences
- 1987 Associate Program Chair, IEEE EMBS International Conference
- 1988 Associate Program Chair, IEEE EMBS International Conference
- 1989 IEEE TAB Committee on International Participation in Administration
- 1989 IEEE TAB Opinion Survey Committee
- 1985–1997 Associate Editor, IEEE Engineering in Medicine and Biology Magazine
- 1987–1991 Member, Nominations Committee, IEEE EMBS
- 1988–1992 International Chairman, IEEE EMB Society
- 1988–1992 IEEE Student Activities Committee
- 1990–1996 Judith Resnik Award Committee for Aerospace Engineering Contributions
- 1990–1995 Internationally elected At-Large Member, Admin Committee of EMBS
- 1990–1991 International Program Chair, IEEE EMBS International Conferences
- 1991–1997 IEEE Distinguished Speakers Pool
- 1992 General Conference Co-Chair, IEEE EMBS International Conference
- 1992–1997 IEEE EMBS Inter Society Liason Committee
- 1992 Delegate to the Public Policy Commission, Am Inst of Med and Biol Eng
- 1992–1995 Delegate to the Council of Societies, Am Institute of Med and Biol Eng
- 1993 International Program Chair, IEEE EMBS International Conference
- 1993 Organizing Committee, EMBS Summer School for Image Processing
- 1993 EMBS Delegate to the European Society for Engineering in Medicine
- 1993–1996 IEEE EMBS Awards Committee
- 1993–1998 IEEE EMBS Publications Committee
- 1992–1995 IEEE Professional Activities Committee
- 1994 Vice President, IEEE Engineering in Medicine and Biology Society
- 1995–1997 Founding Chairman, Regional Conferences Committee
- 1994–1996 Chairman, IEEE EMBS Conference Committee
- 1994–2000 IEEE Committee on Communications and Information Policy
- 1994 Finance Director, World Congress on Computational Intelligence
- 1996 Editor-in-Chief, IEEE Transactions on Info Techn in Biomedicine
- 1996 EMBS Delegate to IEEE Neural Nets Council
- 1997– Co-Chair, Information Technology in Biomedicine Conference, Prague
- 1996– Co-Chair, Information Technology in Biomedicine Conference, DC
- 1996– Co-Chair, Information Technology in Biomedicine Conf, Amsterdam
- 1996– Editorial Board, IEEE Press Series on Biomedical Engineering
- 1996– Co-Chair, Information Technology in Biomedicine Conference, Arlington
- 2001– Editor Emeritus, IEEE Trans on Information Technology in Biomedicine
- 2002– Editorial Board, IEEE Transactions on Neuro Biosciences
- 2002–2004 Elected to the IEEE Publications Services and Products Board
- 2002 IEEE PSPB Delegate to the IEEE Transnational Committee

External Services

Non-IEEE Conference Scientific Advisory Board (Representative Titles):

- 1980 International Conference of Physiological Sciences, Budapest, Hungary
- 1988 14th International Conference of the Indian Biomedical Eng Soc, Bombay
- 1989 International Measurement Confederation Conference on Biomed Instr, India
- 1990 International Measurement Confederation Conf on Clinical Res Models, Hungary
- 1992 International Conference on Biomedical Eng in the 21st Century, Teipei, Taiwan
- 1993 Advanced Technologies and Competitiveness, Electra International, Edison, USA
- 1994 International Conference of the International Federation of Auto Control, Italy
- 1995– Computer-Assisted Radiology International Conferences
- 1995 Frontiers in Medical Visualization, BioMedVis '95, Atlanta, Georgia, USA
- 1995 Baltic International Conference on Biomedical Engineering, Finland
- 1996 11th International Symposium on Biomedical Engineering, Zagreb, Croatia
- 1996– VMW Virtual Magazine on Telemedicine Editorial Advisory Board, Belgium
- 1996 NBC '96, Advisory Committee, Ragnar Granite Institute, Finland
- 1997 International Conference on Computer-Assisted Radiology, Berlin, Germany
- 1998 2nd International Conference on Bioelectromagnetism, Melbourn Australia
- 1998 VIII Mediterranean Conference, MedCon '98, Cyprus
- 1998 8th International IMEKO Conference on Biomedical Measurements, Croatia
- 1999 Nordic-Baltic Conference, Tallinn, Croatia
- 1999 Honor Committee, Health Emergency Telematics, HET '99, EU, Spain
- 2000 International Workshop on Multimedia Applications, Hong Kong
- 2000 International Conference on Computer Assisted Radiology and Surgery, Germany
- 2000 International Conference on Biomedical Engineering, Madras, India
- 2000 MEDICON 2001, Croatia
- 2002 4th IASTED International Conference on Signals and Image Processing, Howaii
- 2003 5th International Federation of Automatic Control Conference, Sydney, Australia

- 2004 Chair, Scientific Advisory Board, International Congress on Medical and Care Compunetics, Den Hague, The Netherlands
- 2005 Chair, Scientific Advisory Board, ICMCC Event 2005, Den Hague, The Netherlands

Academic Committees Served (Representative Titles)

- 1983– Ph.D. Thesis Committees, University of Medicine & Dentistry, Newark, NJ, USA
- 1984– M.S, Thesis Committees, New Jersey Institute of Technology, Newark, NJ, USA
- 1985 Faculty Promotions Review Committee, Indian Institut Science, Bangalore, India
- 1986 Faculty Promotions Review Committee, Technical University of Delft
- 1988 Faculty Promotions Review Committee, Dalhousie University, Nova Scotia
- 1988 Ph.D. Thesis Committee, Indian Institute of Technology, Madras, India
- 1989– Networks Resources Consortium, NJIT, UMDNJ, SIT, New Jersey, USA
- 1990– Ph.D. Thesis Committees, New Jersey Institute of Technology, Newark, NJ, USA
- 1990– University Academic Computing Committee, Univ of Med, Newark, NJ, USA
- 1991– Continuing Education Committee, University of Medicine & Dentistry, Newark
- 1992 Ph.D. Thesis Committee, Indian Institute of Technology, New Delhi, India
- 1993 Campus-Wide Information System Advisory Committee, Univ of Medicine, NJ
- 1993– Strategic Planning Committee, University of Medicine & Dentistry, Newark
- 1994 NJ Equipment Loan Grants Steering Committee, University of Med, NJ, USA
- 1995 World Wide Web Advisory Committee, University of Medicine & Dentistry, NJ
- 1993 Molecular Modeling User Committee, University of Medicine & Dentistry, NJ
- 1997 Technology Policy Committee, NextGen Internet, Princeton
- 1998 NJ Voice Over IP Global Partnership Committee, Fort Lee, NJ
- 2000 Information Technology in Education Committee, NL University, Illinios
- 2001 Presidents Steering Committee, National Louis University, Illinois
- 2001 University Technology Committee, National Louis University, Illinois

Services as Journal Reviewer and Associate Editor

Journal of Medical and Biological Engineering and Computing
IEEE Transactions on Biomedical Engineering
IEEE Engineering in Medicine and Biology Magazine
IEEE Proceedings
Physics in Medicine and Biology
Journal of Basic and Applied Biomedicine
IEEE Transactions on Information Technology in Biomedicine
Critical Reviews in Biomedical Engineering
Applied Journal of Physiology,
Cardiovascular Research
Annals of Biomedical Engineering

Professional Society Memberships (Past to Present)

Royal Aeronautical Society, London, UK
Institute of Electrical Engineers
The Dutch Biophysical Society, The Netherlands
Institute of Electrical and Electronics Engineers, Inc (Senior Member)
IEEE Engineering in Medicine and Biology Society
IEEE Computer Society
IEEE Communications Society
American Association for the Advancement of Science
New York Academy of Science
Biomedical Engineering Society of India (Life Member)
Japanese Biomedical Engineering Society
Romanian Society of Clinical Engineering and Computing (Life Member)
American Medical Informatics Association
Academy of Biomedical Scientists (Life Member)
Biophysical Society (Founding Member)
American Institute of Medical and Biological Engineering (Fellow)
International Council on Medical & Care Compunetics (Founding Member)

List of Publications

Keywords

AIDS Research, Bioinformatics, Biomedical Signals Processing, Biomedical Supercomputing, Biostatistics, Cardiology and Cardiovascular Dynamics, Circadian Physiology, Compunetics, Computer-Aided Medical Devices, DNA Sequence Analysis, Drug Design, Electronic Commerce, Genetic Engineering, Health Care Technology, High Performance Computing, Image Processing, Information Technology, Instructional Technology, Internetworking, Leprosy Research, Mathematical Modeling, Molecular Biology, Molecular Modeling and Dynamics, Networking, Neuroscience, Pharmacology, Physiological Systems Simulation, Protein Engineering, Pulmonary Physiology, Sleep Research, Sudden Infant Death Syndrome, Telecommunications, Telemedicine, Visualization, Voice and Video over IP, Wireless Networking

Networking, Communications and Biomedicine

L. Michelson, S. Laxminarayan (1982), *A distributed mini network for biomedical data processing*, In: proceedings of the 17th Conference of the Association for the Advancement of Medical Instrumentation, 81

L. Michelson, S. Laxminarayan (1982), *Biomedical signals processing in a distributed network environment*, In: Applications of Computers in Medicine, Ed: M.D. Schwartz, IEEE Press, 215–224

S. Laxminarayan (1984), *A biomedical signals processing system*, In: Proceedings, 37th Annual Conference of Engineering in Medicine and Biology, LA, 34, 221

D. Mukhedkhar, S. Laxminarayan et al (1990), *Global telemedicine using INMARSAT satellite system*, SPIE Proceedings, 1355: 19–24

S. Laxminarayan, S. Parmett, J. Yadav et al (1992), *Creating an electronic environment in the workplace*, In: IEEE EMB Magazine, Vol 1, 35–41

S. Laxminarayan, M. Majidi, J. Yadav et al (1992), *Advanced computing and information technology in biomedical research*, In: Journal of Biomedical engineering Applications, Basic and Communication, Vol 2, 108–109

S. Laxminarayan, J. Yadav, M. Majidi et al (1993), *Biomedical computing and information technology*, In: Frontiers of Computing in Biomedical Engineering, Ed: N. Thakor, S. Laxminarayan

G. R. Rao, S. Laxminarayan, B.A. Suresh (1993), *Towards a framework for a decision support system for integrated information retrieval in biomedical computing,* In: IEEE EMBS Proceedings, Ed: R. Rangayyan, A. Szeto, San Diego, CA

S. Laxminarayan, J. Yadav, M. Majidi (1994), *Impact of HPCC initiatives on biomedical research and education*, Journal of Medical and Biological Engineering, 272

S. Laxminarayan, M. Majidi, J. Yadav, R. Fekete (1994), *High Performance Computing and Communications Initiatives: How does it serve the U.S. biomedical engineering community*, In: Proceedings, Third International Conference on Biomedical Engineering in the 21st Century, Ed: Wang, Taiwan

S. Laxminarayan (1995), *Grand Challenges in biomedical sciences and engineering: Role of Information Superhighway*, In: Proceedings, First IEEE EMBS International Regional Conference, Delhi, Ed: S.K. Guha

S. Laxminarayan, M. Majidi, D. Yatcilla (1995), *Advanced computing and communications issues in medicine*, In: Proceedings, IEEE EMBS regional Conference, New Zealand

S. Laxminarayan, G. Rao (1995), *National Information Infrastructure: An information paradigm in biomedicine,* Editorial, In: Journal of Basic and Applied Biomedicine

S. Laxminarayan, M. Majidi, D. Yatcilla, R. Fekete (1995), *NII: The digital nirvana for health and medical research*, In: Proceedings, International Workshop on Mechatronics in Surgery, Bristol, Ed: P. Brett

S. Laxminarayan, P. Yadav (1996), *Biomedical Information technology: Internet and Beyond*, In: Proceedings, IEEE EMBS International Conference, Amsterdam, Ed: W. Rutten, M. Neumann

S. Laxminarayan (1996), *Internet and Health Care*, In: Proceedings, International Symposium on "Past, Present and Future of Internet", San Paulo, Brazil

S. Laxminarayan (1996), *Technological Innovation: Horizon of New Opportunities*, The Telecommunications Summit: Capitalizing on Business Opportunities for New Jersey, Organizer: Congressman Bob Franks, Somerset, NJ

S. Laxminarayan (1996), *High Performance Computing and Communication advances in biomedical applications*, In: Analysis of Biomedical Signals and Images, Ed: J. Jan, P. Kilian, I. Provaznik, BRNO Press, 1–3

S. Laxminarayan (1996), *Information management in medicine*, In: Proceedings, Southern Biomedical Engineering Conference, Dayton

S. Laxminarayan (1996), *Internetworking in medicine: Status and Issues*, In: Proceedings, 3rd Asian-Pacific Conference on Medical and Biological Engineering, Taiwan

S. Laxminarayan (1996), *Advanced computing: Impact on achieving improved quality of life*, In: Proceedings, 3rd Asian-Pacific Conference on Medical and Biological Engineering, Taiwan

S. Laxminarayan (1997), *Biomedical Information Technology: Medicine and Health Care in the Digital Future*, Editorial, IEEE Transactions on Information Technology in Biomedicine, Vol 1, 1–8

S. Laxminarayan, P. Macias (1997), *Impact of Internet on Pharmaceutical and Health Care Industry (Keynote)*, In: Proceedings, IEEE EMBS International

Conference on Information Technology Applications in Biomedicine, Prague, Ed: Ivan Krekule

S. Laxminarayan (1998), *Health Care Information Technology: What is on the Horizon*, IEEE Transactions on Information Technology in Biomedicine, Vol 4

S. Laxminarayan, E. Micheli-Tzanakou (1998), *Information Technology Applications in Biomedicine*, IEEE Press, Washington DC, May, Editors: Laxminarayan and Tzanakou

S. Laxminarayan (1998), *Information Technology in Medicine: The Next Step Beyond*, In: Medical and Biological Engineering and Computing, Supplement 1

S. Laxminarayan (1999), *Biomedical Information Technology: From Digital Dawn to NGI*, 3rd IEEE International Conference on Information Technology Applications in Biomedicine, Amsterdam, Ed: Marsh and Laxminarayan, Elsvier Press

S. Laxminarayan (1998), *Technology Transfer in Biomedicine: Global Criteria and Positioning*, Proceedings of the First International Conference on Health Care Emergency Telematics, Sevilla, Spain

S. Laxminarayan (1999), *Telemedicine and Telemedical Solutions*, Editorial, IEEE Transactions on Information Technology in Biomedicine, Vol 2, 82–83

S. Laxminarayan (1999), *Information Technology Road Map in Biomedicine: Overview*, In: Medical and Biological Engineering and Computing, Vol 37, Supplement 1

S. Laxminarayan (1999), *Biomedicine in the 21st Century: Impact of Information Technology*, IEEE Transactions on Information Technology in Biomedicine, Vol 3, 1, 2–6

S. Laxminarayan (2000), *Emerging Trends at the Threshold of a New Millennium*, IEEE Transactions on Information Technology in Biomedicine, 4, 1, 2–6

R. Istepanian, S. Laxminarayan (2000), *Next Generation of Integrated Mobile and Internet Telemedicine Systems*, Journal of Medical and Biological Engineering Supplement

S. Laxminarayan (2000), *IT Road Map: Concepts, Protocols & Challenges*, Proceedings, The Great Lakes Conference, Wisconsin

S. Laxminarayan, R. Istepanian (2000), *UNWIRED: The Next Generation of Wireless and Internet Telemedicine Systems*, In: IEEE Transactions on Information Technology in Biomedicine, Vol 4, 189–193

S. Laxminarayan (2000), *Emerging Trends in Biomedical Information Technology*, Journal of Medical and Biological Engineering Supplement

S. Laxminarayan (2001), *Cross Roads in Global Tele-Issues*, Journal of Medical and Biological Engineering Supplement, IFMBE Proceedings, 13–15

S. Laxminarayan (2001), *How Health Care and Engineering Blend in the IP Universe*, In Proceedings, ICBME Conference Madras, India

S. Laxminarayan (2001), *Barriers and Bounds in Internet Medicine*, In Proceedings of the Eighth Australian and New Zealand Intelligent Information Systems Conference, Perth, Australia, 2001

S. Laxminarayan (2002), *Information Technology in Biomedicine: Maturational Insights*, IEEE Transactions on Information Technology in Biomedicine, Vol 6, Number 1, 1–7

S. Laxminarayan, B.H. Stamm (2002), *Technology, Telemedicine and Telehealth:* Business Briefing - Global Healthcare 2002 vol. 2. 93=97, World Medical Association., London, UK

S. Laxminarayan, L. Kun (2002), *Combating Bioterrorism with Bioengineering*, IEEE Engineering in Medicine and Biology Magazine, Vol 21, 21–27

B.H. Stamm, S. Weeg, S. Khabir, S. Laxminarayan et al (2002), *Going the Extra Mile: Macro and Micro Issues in Implementing Telehealth and Telemedicine*, In Proc, Rural Telecommunications Congress, Des Moines, IA

J.S. Suri, S. Laxminarayan (2002), *Angiography and Plaque Tissue Imaging: Computer and Image Analysis Techniques*, In: BEACON Annual Symposium Proceedings, Hartford, Connecticut

S. Laxminarayan (2003), *The Obtrusive Dimensions to Bioterrorism*, In: BEACON Annual Symposium Proceedings, Hartford, Connecticut

S. Khabir, S. Laxminarayan, B.H. Stamm (2003), *Evaluating Potential Telehealth Activities on Different Types of Bandwidth*, In: Proceedings American Telemedicine Association Conference. (Selected as one of the best posters by the American Telemedicine Association)

S. Laxminarayan, B.H. Stamm (2003), *Issues and Technological Challenges in Counter Bioterrorism*, In: Proceedings, HealthCom 2003

S. Laxminarayan, B.H. Stamm (2003), *Information Engineering in Medicine and Healthcare: Hype, Reality and the Future*, In Proc, International Conference on Medical Informatics and Engineering, Craiova, Romania

S. Laxminarayan (2003), *Next Generation Internet Initiative: Video Streams and Distance Learning using VoIP technologies*, In Preparation

S. Laxminarayan (2003), *Electronic Commerce Applications in Health Care*, In Preparation

Bioinformatics, Genetic and Protein Engineering

J. Yadav, S. Laxminarayan (1990), *Molecular modeling in AIDS research: Development of a receptor map for the active site of HIV reverse transcriptase as an aid in the design of inhibitors*, In: Proceedings, International Biophysics Conference, 2, 257

J. Yadav, S. Laxminarayan et al (1990), *Molecular modeling of reverse transcriptase inhibitors: Implications in the treatment of AIDS*, In: Proceedings, IEEE EMBS Conference, Philadelphia, Ed: P. Pederson & B. Onaral, Vol 12, 1606–1607

J. Yadav, P. N. Yadav, E. Arnold, S. Laxminarayan, M. Modak (1990), *Molecular dynamics studies of polymerase binding site of klenow fragement of DNA polymerase* I, In: Journal of Biomolecular Stereodynamics

J. Yadav, P. N. Yadav, E. Arnold, S. Laxminarayan, M. Modak (1990), *Molecular modeling of antiviral agents for AIDS virus*, In: Proceedings, Workshop on Computational Aspects of Inorganic Chemistry in Biological Systems

J. Yadav, S, Laxminarayan (1991), *Computer assisted design of antiviral agents directed against Human Immunodeficiency Virus reverse transcriptase as their target*, In: Annals of the New York Academy of Sciences, Vol 616, 624–630

J. Yadav, P. N. Yadav, E. Arnold, S. Laxminarayan, M. Modak (1991), *Molecular modeling of the interactions between Escherichia coli DNA Polymerase I and substrates*, In: Proteins: Structure, Dynamics, Design, Ed: V. Renugopalakrishnan, P. R. Carey, I. C. P. Smith, S. G. Huang, A. C. Storer, ESCOM, Leiden, 330–335

J. Yadav, S. Laxminarayan et al (1991), *Molecular modeling of structural mimicry of potential antiviral agents*, In Proceedings, World Organization of Theoretical Organic Chemists

P. N. Yadav, J. Yadav, C. Wong, S. Laxminarayan (1991), *Computer assisted design of purine nucleoside analogs: Implications in AIDS therapy*, In: IEEE EMBS Proceedings, Orlando, Ed: J. Nagel & W. Smith, Vol 13, 1487–1488

D. S. Kristol, M. A. Fisher, J. Yadav, S. Laxminarayan (1992), *Steric effects at acyclic and bridgehead carbons attached to carbonyls*, In: IEEE EMBS Proceedings, Paris, Ed: J. P. Morucci, R. Plonsey, J. L. Coatrieux, S. Laxminarayan, Vol 14, 211–212

A. Ritter, C. M. Gerula, J. Yadav, S. Laxminarayan (1992), *Active analog approach to histamine H1 receptor antagonist*, In: IEEE EMBS Proceedings, Paris, Ed: J. P. Morucci, R. Plonsey, J. L. Coatrieux, S. Laxminarayan, Vol 14, 208–210

J. Yadav, S. Laxminarayan (1993), *The role of visualization in protein engineering*, In: Advanced Technologies and Competitiveness, Electro '93 International, Vol 3, 414–417

M. A. Fisher, P. N. S. Yadav, J. Yadav, S. Laxminarayan (1993), *A computer assisted receptor mapping approach to the design of anti-AIDS agents directed at HIV reverse transcriptase*, In: Proceedings, 19th North East Bioengineering Conference, Ed: J. Li, Vol 19, 162–163

N. Wokhlu, J. Yadav, A. Ritter, D. Kristol, S. Laxminarayan (1995), *Identification of a pharmacophore on H1 histamine receptor antagonists: A molecular modeling study*, In: Proceedings, IEEE EMBS Regional Conference on Mass Health Care, Delhi

J. Yadav, S. Laxminarayan, E. Arnold, P. Yadav, M. Modak (1994), *Molecular modeling studies of purine dideoxynucleosides aimed at the improved design of potent anti-HIV inhibitors*, In: Journal of Molecular Structures and Theoretical Chemistry

P.N.S. Yadav, S. Laxminarayan (1996), Three dimensional molecular modeling studies in AIDS research using high performance computing facilities: A ternary complex of HIV-1 RT, In IEEE EMBS Proceedings, Amsterdam, Ed: Wim Rutten, M. Neumann, H.B.K. Boom

P.N.S. Yadav, S. Laxminarayan (2000), *Enzymatic mechanism of Action of AIDS Drugs: A Molecular Simulation Study*, Journal of Medical and Biological Engineering Supplement

Biomedical Supercomputing

S. Laxminarayan (1988), *Application of supercomputers in biomedical research*, Guest Editorial, IEEE Engineering in Medicine and Biology Magazine, Vol 7, No. 4, 11

S. Laxminarayan, L. Michelson (1988), *Perspectives in biomedical supercomputing*, IEEE Engineering in Medicine and Biology Magazine, Vol. 7, 4, 12–15

S. Laxminarayan (1988), Guest Editor, *Supercomputing*, IEEE EMB Magazine, 7, 4, 11–43. Supercomputing in Medical Imaging, Supercomputer applications in molecular modeling, Supercomputer applications to DNA sequence analysis, Large scale simulations of the hippocampus, Supercomputer use in orthopaedic biomechanics research

S. Laxminarayan (1989), *Role of supercomputers in medicine and biology*, In: Proceedings, IMEKO 5th International Conference on Biomedical Instrumentation, Calcutta, India, Ed: S. Chatterjee

J. Yadav, S. Laxminarayan, L. Michelson (1989), *Supercomputers in drug design*, Proceedings, IEEE EMBS Conference, Vol 11, 1901–1902, Ed: F. Spellman & Y. Kim

S. Laxminarayan, L. Michelson (1989), *Perspectives in biomedical supercomputing*, In: Technology Monitor, IEEE Engineering Management Review, Vol 17, 4

W. C. Lambert, R. Trost, S. J. Robboy, S. Laxminarayan (1989), *A supercomputer based system for complex multivariate analysis of dermatopathologic specimens*

S. Laxminarayan (1990), *Supercomputing, visualization and physiological models*, In: Proceedings, 11th Annual Conference of the Chinese Society of Biomedical Engineering, Nanking, China

J. Yadav, S. Laxminarayan, E. Arnold, P. Yadav, M. Modak (1990), *Supercomputer applications in the design of antiviral agents for AIDS treatment*, In: Proceedings, IEEE EMBS Conference, Vol 11, 1606–1607, Ed: P. Pederson & B. Onaral

K. Krishnan, S. Laxminarayan, T. Terry (1991), *Advances in biomedical supercomputing*, In: Proceedings, IEEE EMBS Conference, Vol 13, 1228–1229, Ed: J. Nagel & W. Smith

Signals Processing in Medicine & Biology

S. Laxminarayan, W.J.A. Goedhard et al (1976), *Analysis of the human acceleration ballistocardiogram*, In: Non-Invasive Mechanical Methods in Cardiology and Cardiovascular Dynamics, Ed: W.J.A. Goedhard, S. Karger-Basel, NY, 246–250

S. Laxminarayan, A.J.G. Spoelstra et al (1977), *Digital filtering of transpulmonary pressure in lung compliance studies*, In: Proceedings, 18th Dutch Federation Meeting, Leiden, 277

S. Laxminarayan, A.J.G. Spoelstra et al (1978), *Transpulmonary pressure and lung volume of the cat and the newborn: Removal of cardiac effects with a digital filter,* In: Journal of Medical and Biological Engineering and Computing, 16, 397–407

S. Laxminarayan, A.v.d. Vos et al (1978), *Windowing method of digital filtering: Applications in physiology*, In: Proceedings, DECUS Symposium, Copenhagen

S. Laxminarayan, S. Chatterjee, E.D. Weitzman et al (1980), *Spectral analysis of sleep stages in infants and adults,* In: Proceedings, 20th Meeting of the Association for the psychophysiological study of sleep, 187

S. Laxminarayan, O. Mills, A.C. Cornwell et al (1980), *Convergence behavior and resolution criteria of Walsh and Haar transforms in the analysis of sleep stage patterns*, In: Proceedings, International Symposium on Mathematical and Computational Methods in Physiology, International Congress of Physiological Sciences, Budapest, Hungary

J. Zimmerman, C.A. Czeisler, S. Laxminarayan et al (1980), *Eye movement density during REM sleep in normal 'free-running' human subjects*, In: Proceedings, 20th Meeting of the Association for the Psychophysiological Study of Sleep, Mexico, 133

J. Zimmerman, C.A. Czeisler, S. Laxminarayan et al (1980), *REM density is disassociated from REM sleep during 'free-running' sleep episodes*, In: Sleep, 2,4, 409–415

S. Laxminarayan, O. Mills et al (1981), *Haar transform applications in sleep apnea studies*, In: Proceedings, IEEE Frontiers of Engineering and Computing in Health Care, Ed: B. Cohen, 127–131

S. Laxminarayan, O. Mills et al (1981), *Haar transform applications in sleep studies*, In: IEEE Transactions on Biomedical Engineering, 8

S. Laxminarayan, L. Michelson, J.J. McArdle et al (1981), *Real time processing of electrophysiological events at the neuromuscular junction*, In: Biomaterials, Medical Devices and Artificial Organs, 4, 257–258

R. Laxminarayan, L. Rajaram, S. Laxminarayan (1981), *Online digital filtering applications in physiology*, In: Biomaterials, Medical Devices and Artificial Organs, 4, 260–261

R. Laxminarayan, L. Rajaram, S. Laxminarayan (1982), *Online digital filter applications using windowing techniques*, In: Biomedical Engineering: Recent Developments, Ed: S. Saha, Pergamon Press, 60–64

S. Laxminarayan, L. Rajaram, A.C. Cornwell et al (1982), *Spectral analysis applications in the characterization of sleep-wake patterns of normal and "near miss" infants for the SIDS*, In: Proceedings, 10th Annual Northeast Bioengineering Conference, IEEE, 145–150

M. Markowitz, J.F. Rosen, S. Laxminarayan et al (1982), *Periodic oscillations in serum 1.25 DIHYDROXIVITAMIN D in humans*, In: Clinical Research, 30, 2, 399

M.E. Markowitz, J.F. Rosen, S. Laxminarayan et al (1982), *Serum parathyroid (PTH) concentrations may undergo rhythmic changes in humans*, In: Clinical Research, 30, 3, 700

M.E. Markowitz, J.F. Rosen, S. Laxminarayan et al (1984), *Circadian rhythms of blood minerals during adolescence*, In: Pediatric Research, 5, 18, 456–462

S. Laxminarayan, O. Mills et al (1985), *Spectral analysis of electrophysiological events in SIDS*, In: Proceedings, 11th Northeast Bioengineering Conference, IEEE 114–119

S. Laxminarayan, A.C. Cornwell et al (1986), *Spectral analysis applications in infant sleep and respiratory data analysis*, In: Proceedings, IEEE EMBS International Conference, Ed: C.R. Robinson & G. Kondraske, 8, 1195–1196

A.C. Cornwell, S. Laxminarayan, M. Feuerman (1987), *Sleep loss affects apnea in near-miss SIDS, but not control babies*, In: Medical Electronics, 9, 95

A.C. Cornwell, S. Laxminarayan et al (1988), *Developmental changes in electrophysiological and spectral characteristics of sleep apnea in infants following a Sudden A-Ventilatory Event (S.A.V.E.) and controls,* In: Proceedings, Conference on SIDS: A look into future, Charleston, South Carolina

M.E. Markowitz, J.F. Rosen, S. Arnaud, S. Laxminarayan (1988), *Temporal interrelationships between the circadian rhythms of serum parathyroid harmone and calcium concentrations*, In: Journal of Clinical Endocronology and Metabolism

S. Laxminarayan, L. Michelson, A.C. Cornwell (1988), *Bispectral Analysis*, In: Computers in Health Sciences, Ed: S. Haque, Newark, New Jersey

S. Laxminarayan (1988), *Signals processing applications in Sudden Infant Death Syndrome studies*, In: Proceedings, Indian Biomedical Engineering Society Conference, Bombay, India

A.C. Cornwell, S. Laxminarayan (1989), *A polysomnographic sleep abnormality in Sudden A-Ventilatory Event (S.A.V.E.) at high risk for SIDS infants*, In: Proceedings, IEEE EMBS International Conference, Ed: F. Spellman, Y. Kim, Seattle, 11, 313–315

D. Mukhedkar, S. Laxminarayan (1990), *A personal low cost EKG analyzer using fuzzy techniques*, In: Proceedings, IEEE EMBS International Conference, Ed: P. Pederson, B. Onaral, 12

A. C. Cornwell, S. Laxminarayan (1990), *Respiratory dynamics in SIDS*, In: Proceedings, 6th International Conference of the International Measurements Confederation, TC13, Hungary

S. Wiener, S. Laxminarayan, M. Shaikh et al (1990), *Naloxone effects upon hypothalmically elicited masseteric EMG activity*, In: Journal of Dental Research, 69:149

J. D. Prasad, G.J. Huang, S. Laxminarayan, J.J. McArdle (1991), *Fluctuation analysis of calcium channels of murine hippocampal neurons*, In: Proceedings, 35th Annual Meeting of the Biophysical Society, San Francisco

J.D. Prasad, S. Laxminarayan, J.J. McArdle (1991), *Review of data analysis methodologies in ionic channel dynamics*, In: Proceedings, IEEE EMBS International Conference, Ed: J. Nagel and W. Smith, 13, 1530–1531

Computer-aided Physiological Monitoring & Measurement

S. Laxminarayan, E. D. Weitzman et al (1979), *Automatic detection and processing of respiratory pauses in infants*, In: Proceedings, First International Conference on Physiological Measurements, Oxford, UK, Ed: P. Rolfe

S. Laxminarayan, E. D. Weitzman et al (1980), *Walsh Transform applications in the automatic processing of apneas*, In: Foetal and Neonatal Physiological Measurements, Ed: P. Rolfe, Pitman Medical, UK, 266–276

S. Laxminarayan, O. Mills, E. D. Weitzman et al (1980), *Walsh Transform based online respiratory monitoring*, In: Proceedings, International Conference on Recent Advances in Biomedical Engineering, London

S. Laxminarayan, E. D. Weitzman et al (1980), *Online respiratory monitoring in SIDS*, In: IEEE Frontiers of Engineering and Computing in Health Care, 21–24

S. Laxminarayan, A. C. Cornwell et al (1980), *Respiratory monitoring using a mini computer*, In: IEEE Transactions on Biomedical Engineering, Abs, 9, 530

S. Laxminarayan, O. Mills et al (1982), *A digital computer application in the online monitoring of expired carbon-di-oxide, In: Application of Computers in Medicine*, Ed: M. D. Schwartz, IEEE Press, 117–127

S. Laxminarayan, J. J. McArdle et al (1982), *Online analysis of electrical signals from skeletal and cardiac muscle cells*, Medical Computer Science and Computational Medicine, IEEE Computer Society, 539–541

S. Laxminarayan, J. J. McArdle et al (1982), *A digital computer application in the processing of electrophysiological events at the neuromuscular junction*, In: Biomedical Engineering and Recent Developments, Ed: S. Saha, Pergamon Press, 51–55

S. Laxminarayan, O. Mills et al (1983), *Sudden Infant Death Syndrome: A digital computer based apnea monitor*, In: Journal of Medical & Biological Engineering & Computing, 21, 191–196

S. Laxminarayan (1984), *A critical review of digital computer based apnea monitors*, In: Proceedings, 37th Annual Conference of Engineering in Medicine and Biology, Los Angeles, CA, 76

A. Marmarou, A. C. Cornwell, S. Laxminarayan (1985), *Home monitoring of infants at risk for SIDS*, In: IEEE Frontiers of Engineering and Computing in Health Care, Chicago, Ed: B. Feinberg

Mathematical Modeling of Physiological Systems

S. Laxminarayan (1970), *Random process analysis techniques*, In: Proceedings, Conference on Online Computing in Medicine and Biology, Borough Polytechnic, London, 4

A. C. Arntzenius, S. Laxminarayan, H. R. Kulburtus (1971), *Use of computer model in the understanding of the relationship between cardiac excitation and QRS complex*, In: The Electrical Field of the Heart, International Congress of the Physiological Sciences, 21–26

S. Laxminarayan, W. J. A. Goedhard (1976), *A stochaistic model in the analysis of the human acceleration ballistocardiography*, In: Non-Invasive Mechanical Methods in Cardiology and Cardiovascular Dynamics, S. Karger-Basel, NY, 246–250

S. Laxminarayan (1977), *Methods and applications of random data analysis techniques in cardiovascular and pulmonary physiology*, In: Proceedings, International Conference on Theoretical and Applied Mechanics, Surat, 51–53

A. Nagpal, S. Laxminarayan et al (1990), *Mathematical modeling of the immune system in AIDS research*, In: International Journal of the Innovation et Technologie en Biologie et Medecine (ITBM), 11, 5, 519–532

J. D. Prasad, S. Laxminarayan, et al (1990), *Computer simulation of hippocampal neurons*, In: Proceedings, IEEE EMBS Conference, Philadelphia, Ed: P. Pederson & B. Onaral, Vol 12, 1190–1191

A. Nagpal, S. Laxminarayan, et al (1990), *Dynamics of AIDS manifestation: A mathematical prediction*, In: Proceedings, IEEE EMBS Conference, Philadelphia, Ed: P. Pederson & B. Onaral, Vol 12, 1196–1197

Physiological Systems Analysis

P.G. Versteeg, G. Elzinga, S. Laxminarayan (1974), *The influence of changing physical load on aortic blood flow*, In: Arch Int Physiol Biochem, 82(2), 376–381

S. Laxminarayan, P. Sipkema, N. Westerhof (1976), *Impulse response function of the arterial system*, In Proceedings, 17th Dutch Federation Meeting, Amsterdam, 272

S. Laxminarayan, R. Laxminarayan, A.J. Jongbloed (1978), *Application of linear systems analysis by deconvolution techniques in cardiovascular dynamics*, DECUS Symposium Proceedings, Copenhagen

N. Westerhof, G.C. Van Den Bos, S. Laxminarayan (1978), *Arterial reflection*, In: The Arterial System, Ed: Bouer and Busse, Springer Verlag, Berlin, NY, 48–62

S. Laxminarayan, P. Sipkema, N. Westerhof (1978), *Characterization of the arterial system in time domain*, In: IEEE Transactions on Biomedical Engineering, 2, 177–185

S. Laxminarayan, R. Laxminarayan (1978), *Use of swept sine wave in physiological systems analysis*, In: IEEE Transactions on Biomedical Engineering, 1, 103–105

S. Laxminarayan (1978), *Calculation of forward and backward waves in the arterial system*, Journal of Medical and Biological Engineering and Computing, 17, 130

S. Laxminarayan, R. Laxminarayan et al (1979), *Computing total arterial compliance of the arterial system from its input impedance*, In: Journal of Medical and Biological Engineering and Computing, 17, 623–628

S. Laxminarayan, R. Laxminarayan et al (1980), *Arterial Hemodynamics*, In: Proceedings, 2nd Mid-Atlantic Conference on Biofluid Mechanics, Blacksberg

S. Laxminarayan, R. Laxminarayan et al (1980), *Linear systems analysis applications in the study of arterial hemodynamics*, In: Biofluid Mechanics, Ed: D. Schenk, Plenum Publishers, NY, 2, 343–368

R. Laxminarayan, L. Rajaram, S. Laxminarayan (1982), *Estimation of the impulse response function by deconvolution techniques in the characterization of physiological systems*, In: Proceedings, 17th Conference of the Association for the Advancement of Medical Instrumentation, San Francisco, 9

T. Argentieri, J.J. McArdle, S. Laxminarayan (1982), *The sensitivity of the regenerating nerve endings to calcium and 3,4 diaminopyridine*, Neuroscience Meeting, Abstracts

S. Laxminarayan (1983), *Psychophysiological aspects of sleep*, IEEE Proceedings, 72, 4, 542–543

S. Laxminarayan (1983), *Psychophysiological aspects of sleep*, IEEE Engineering in Medicine and Biology Magazine, 56–57

S. Laxminarayan, O. Mills, E.D. Weitzman et al (1985), *Sleep stage and pattern analysis*, Journal of Medical and Biological Engineering and Computing, 505–507

S. Laxminarayan, A.C. Cornwell et al (1985), *Developing characteristic sleep-apnea profiles in SIDS*, In: Frontiers of Engineering and Computing in Health Care, Ed: B. feinberg, 1084–1087

S. Laxminarayan, P. Kadam, S. Gabbay (1986), *Design of a pulse duplicator in heart valve studies*, In: Proceedings, IEEE EMBS International Conference, Ed: C. Robinson, G. Kondraske, 8, 120–124

J. Arena, J.J. McArdle, S. Laxminarayan (1987), *Characterization of the class I antiarrhythmic activity of cibenzoline succinate in guinea pig papillary muscle*, In: Journal of Pharmacology and Experimental Therapeutics, 240, 2, 441–450

A.C. Cornwell, S. Laxminarayan, *EEG and infant sleep*, In: Proceedings, IEEE EMBS International Conference, Ed: K. Foster, R. Neubower, S. Laxminarayan, Vol 9

T.M. Argentieri, S. Laxminarayan, J.J. McArdle (1992), *Characteristics of synaptic transmission in reinnervating rat skeletal muscle*, In: European Journal of Physiology, 421, 256–261

S. Laxminarayan (1994), *Application of a transient analysis method in the characterization of physiological systems*. In: Proceedings, Workshop on Arterial Systems Analysis, Newark, NJ

Visualization and Imaging

S. Ayyadurai, S. Laxminarayan (1989), *Visualization in medicine and biology*, In: Proceedings, IEEE EMBS International Conference, Ed: F. Spellman and Y. Kim, 11

A. Dastane, T.K. Vaidyanathan, S. Laxminarayan (1990), *3D computer generation of occlusal tooth surface*, In: Proceedings, IEEE EMBS International Conference, Ed: P. Pederson, B. Onaral, 12, 1182–1183

T.K. Vaidyanathan, K. Vaidyanathan, S. Laxminarayan (1990), *Digital imaging applications for the evaluation of microstructures of dental biomaterials*, In: Proceedings, IEEE EMBS International Conference, Ed: P. Pederson and B. Onaral, 12

S. Parmett, S. Laxminarayan et al (1991), *Three dimensional visualization program for medical imaging in clinical medicine and education*, In: Proceedings, IEEE EMBS International Conference, Ed: Nagel and Smith, 13

S. Laxminarayan (1991), *3-D visualization in radiological structures*, In: Radiological Society of North America Meeting, Chicago

J. Yadav, S. Laxminarayan (1993), *The role of visualization in protein modeling*, In: Advanced Technologies, ELECTRA '93, 3, 414–417

J. Suri, et al, S. Laxminarayan (2002), *Shape Recovery Algorithms Using Level Sets in 2-D/3-D Medical Imagery: A State-of-the-Art Review*, In: IEEE Trans. in Information Tech in Biomedicine, Vol 6, 1, 8–29

J. Suri, et al, S. Laxminarayan (2001), *Modeling Segmentation via Geometric Deformable Regularizers, PDE and Level Sets in Still and Motion Imagery: A Revisit*, In: Int. Journal of Image and Graphics, 1, 4, 681–734

J. Suri, et al, S. Laxminarayan (2001), *A Revisit of Skeleton vs. Non-Skeleton Approaches for MR and CT Vasculature Segmentation*, In: Annals of Vascular Surgery, In Press

J. Suri, et al, S. Laxminarayan (2002), *Review on real-time Magnetic Resonance Gad-Enhanced Breast Lesion Characterization*, Under Review, In: IEEE Trans. in Information Tech. in Biomedicine

J. Suri, et al, S. Laxminarayan (2002), *A Review on MR Vascular Image Processing Algorithms: Acquisition and Prefiltering: Part 1*, In: IEEE Trans. on Information Tech. in Biomedicine, Vol 4, 324–337

J. Suri, et al, S. Laxminarayan (2002), *A Review on MR Vascular Image Processing: Skeleton Versus Nonskeleton Approaches: Part 2*, In: IEEE Trans. on Information Tech. in Biomedicine, Vol 4, 338–350

J. S. Suri, S. Singh, S. Laxminarayan (2002), *Geometric Regularizers for Level Sets*, In: PDE and Level Sets, Klouwer Press, 97–152

J. Suri, et al, S. Laxminarayan (2002), *White and Black Blood Volumetric Angiographic Filtering: Ellipsoidal Scale-Space Approach*, In: IEEE Trans on Information Technology in Biomedicine, 2, 142–158

J. Suri, et al, S. Laxminarayan (2002), *Vessel Surface Topology Extraction from Noisy White and Black Blood Angiography Volumes Using Scale-Space Framework.* IASTED, Angiography Workshop, SIP 2002

J. Suri, et al, S. Laxminarayan (2002), *Artery-Vein Detection in Very Noisy TOF Angiographic Volumes Using Dynamic Feedback Scale-Space Ellipsoidal Filtering*, IASTED, Angiography Workshop, SIP 2002

X. Shen, J. S. Suri, S. Laxminarayan (2002), *Basics of PDEs and Level Sets*, In: PDE and Level Sets, Klouwer Press, 1–30

J. Suri, et al, S. Laxminarayan (2002), *Contrast Uptake, FCM, Markov Random Field & Live Wire for Lesion Detection in Real-Time MR Breast Perfusion Sequences*

J. Suri, et al, S. Laxminarayan (2002), *A Comparison of State-of-the-Art Diffusion Imaging Techniques for Smoothing Medical/Non-Medical Image Data*, In: Proc. 15th International Conference on Pattern Recognition (ICPR'02), Quebec, 11–15 August, 2002.

J.S. Suri, S. Laxminarayan, J. Gao, L. Reden (2002), *Image Segmentation via PDEs*, In: PDE and Level Sets, Klouwer Press, 153–223

J.S. Suri, S. Singh, S. Laxminarayan (2002), *Medical Image Segmentation Using Level Sets*, In: PDE and Level Sets, Klouwer Press, 301–340

J. Suri, et al, S. Laxminarayan (2002), *Automatic Local Effect of Window/Level on 3-D Scale-Space Ellipsoidal Filtering on Run-Off-Arteries from White Blood Contrast-Enhanced Magnetic Resonance Angiography*, In: Proc. 15th International Conference on Pattern Recognition (ICPR'02), Quebec, (11–15 August, 2002).

J.S. Suri, S. Laxminarayan (2002), *Angiography and Plaque Imaging: Computer and Image Analysis Techniques*, BEACON Annual Symposium Proceedings, Hartford, Connecticut

J.S. Suri, D. Chopp, A. Sarti, S. Laxminarayan (2002), *The Future of PDEs and Level Sets*, In: PDE and Level Sets, Klouwer Press, 385–407

J. Suri, et al, S. Laxminarayan (2002), *An algorithm for Time-of-Flight Black-Blood Vessel Detection*

Time Series Analysis & Stochastic Modeling

S. Laxminarayan, W. J. A. Goedhard (1975), *Time varying mean square measurements of the human acceleration ballistocardiogram*, In: Proceedings, 4th World Congress on Ballistocardiography and Cardiovascular Dynamics, Amsterdam, Ed: W. J. A. Goedhard, 64

S. Laxminarayan, W. J. A. Goedhard (1975), *Uncovering the state of cardiac activity from acceleration BCG records using non-stationary analysis techniques*, In: Proceedings, 4th World Congress on Ballistocardiography and Cardiovascular Dynamics, Amsterdam, 65, Ed: W. J. A. Goedhard

A. C. Cornwell, E. D. Weitzman, S. Laxminarayan (1980), *Sleep onset REM periods in infants*, In: Proceedings, 20th Meeting of the Association for the Psychophysiological Study of Sleep, Mexico, 173

A. C. Cornwell, S. Laxminarayan (1986), *Effects of sleep deprivation on apnea in near-miss SIDS infants*, IEEE EMBS Conference, Fort Worth, Texas, Vol 8, 1183–1187, Ed: C.R. Robinson & G. Kondraske

A. C. Cornwell, S. Laxminarayan (1987), *Sleep apnea in near-miss and control infants*, In: Proceedings, IEEE EMBS Conference, Boston, Vol 9, Ed: K. Foster, R. Newbower, S. Laxminarayan

A. C. Cornwell, S. Laxminarayan (1987), *Sleep apnea characteristics of infants with a Sudden A-Ventilatory Event (S.A.V.E.) and controls*, In: Pediatric Pulmonology, Nov-Dec Issue

A. C. Cornwell, S. Laxminarayan (1987), *Sleep apnea in S.A.V.E. infants*, In: proceedings, 6th Annual Apnea of Infancy Conference, Annenberg

A. C. Cornwell, S. Laxminarayan (1988), *Biological rhythms of apnea in relation to sleep deprivation in S.A.V.E. and control infants*, In: Proceedings, IEEE EMBS Conference, Vol 10, 1803–1804, Ed: G. Harris, C. J. Walker

A. C. Cornwell, S. Laxminarayan (1993), *A sleep disturbance in high risk for SIDS infants*, In: Journal of Sleep Research, 2, 110–114

Computers in Education

S. Laxminarayan (1986), *A biomedical signals processing system for medical education*, In: Proceedings, IEEE EMBS Conference, Fortworth, Texas, Vol 8, Ed: C. R. Robinson & G. Kondraske

S. Laxminarayan, J. Yadav, S. Dunn, A. Ritter et al (1990), *Computer curricula in a biomedical sciences graduate environment*, In; Proceedings, IEEE EMBS Conference, Vol 12, Ed: P. Pederson & B. Onaral

S. Laxminarayan, M. Majidi, J. Yadav et al (1990), *Impact of computers in medical education*, In: Proceedings, International Conference on Medical Education, Prague

S. Parmett, S. Laxminarayan, M. Majidi et al (1991), *Three dimensional visualization of medical imaging data: Educational Applications*, In: Proceedings, Meeting of the Radiological Society of North America

S. Laxminarayan, K. Krishnan, T. Terry (1991), *Shared resources for biomedical education and research*, In: Proceedings, IEEE EMBS Conference, Vol 13, Ed: J. Nagel & W. Smith

S. Laxminarayan, J. Bronzino, J. W. Beneken et al (1994), *The role of professional societies in biomedical engineering*, In: Biomedical Engineering Handbook, CRC Press, Ed: J. Bronzino, 2787–2793

S. Laxminarayan, J. Bronzino et al (1994), *Professional societies in BME*, In: proceedings, Third International Conference on Biomedical Engineering in the 21st Century. Ed: Wang

Health Care Technology

T. G. Krishna Murthy, S. Laxminarayan (1986), *Information technology in third world countries*, In; proceedings, IEEE EMBS Conference, Vol 8, Ed: C. R. Robinson & G. Kondraske, 1715–1718

S. Laxminarayan, T. G. Krishna Murthy (1986), *Health care technology in the third world: Man power needs and training*, In: Proceedings, 12th CMBEC Meeting and First International Pan-Pacific Symposium, Vancouver, 186–187

D. Mukhedkar, S. Laxminarayan et al (1990), *Resource allocation for health care: Philosophy, Ethics and Law*, In: proceedings, International Conference on Telecommunications for Health Care: Telemetry, Teleradiology and Telemedicine, Calgary

V. G. Rao, S. K. Sahu, S. Laxminarayan (1992), *Health services system - How effective?* In: Proceedings, IEEE EMBS Conference, Ed: Morucci, Plonsey, Coatrieux, Laxminarayan, Paris, 1173–1174

V. Thevenin, J. L. Coatrieux, S. Laxminarayan (1993), *Biomedical engineering around the world*, Medical Technology Symposium of the European Economic Community, In: IEEE Engineering in Medicine and Biology Magazine, 12, 2, 42–63, The European scene of medical technology, J. E. W. Beneken

BME Research in Europe: A View from the industrial standpoint, A. Oppelt BME strategies in the U.S., R. Nerem Research Strategies in Japan, F. Kajiya

Last Publications (2004–2005)

S. Laxminarayan, L. Kun (2004), *The Many Facets of Homeland Security*, In: IEEE Engineering in Medicine and Biology Magazine, 23,1,19–29

S. Laxminarayan (2004), *The Knowledge Management Paradigm*, In Telehealth Practice Report, Nov/Dec 9(5): 1,6

S. Laxminarayan, B.H. Stamm (2004), *Issues and Technological Challenges in Counter Bioterrorism*, IEEE Engineering in Medicine and Biology Magazine, 23, 1, 119–121

S. Laxminarayan, B.H. Stamm, N. Piland, S. Weeg (2004), *Store & Forward Technology in Telemedicine*, In: Telemedicine Journal and e-Health, Suppl 1, S43

L. Bos, S. Laxminarayan, A. Marsh (2004), *Healthcare Compunetics*, In: Medical and Care Compunetics 1, IOS Press, 3–11

N. S. G. Raju, J. P. Anthony, R.S. Jamadagni, R. Istepanian, J. Suri, S. Laxminarayan (2004), *Status of Mobile Computing in Healthcare: An Evidence Study*, In IEEE Engineering in Medicine and Biology Proceedings

R. Istepanian, N. Philip, X.H. Wang, S. Laxminarayan (2004), *Non-Telephone Healthcare: The Role of 4G and Emerging Mobile Systems for Future m-Health Systems*, In: Medical and Care Compunetics 1, IOS Press, 465–470

R. M. Oberleitner, S. Laxminarayan, (2004), *Information Technology and Behavioral Medicine: Impact on Autism Treatment and Research*, In: Medical and Care Compunetics 1, IOS Press, 215–222

R. M. Oberleitner, S. Laxminarayan, J. Suri, (2004), *The Potential of a Store and Forward Tele-Behavioral Platform for Effective Treatment and Research of Autism*, In: IEEE Engineering in Medicine and Biology Proceedings, 3294–6

S. Laxminarayan (2005), *Invited Foreword*, In: Clinical Knowledge Management: Opportunities and Challenges, Rajeev K. Bali, Editor, Idea Group, UK, viii–x

J. S. Suri, C. S. Pattichis, C. Li, J. Macione, Z. Yang, M. D. Fox, D. Wu, S. Laxminarayan (2005), *Plaque Imaging Using Ultrasound, Magnetic Resonance and Computer Tomography: A Review*, In: Plaque Imaging, Pixel to Molecular Level, IOS Press, 1–25

B. Tulu, S. Chatterjee, S. Laxminarayan (2005), *A Taxonomy of Telemedicine Efforts with respect to Applications, Infrastructure, Delivery Tools, Type of Setting and Purpose*, Annual Hawaii International Conference on System Sciences, Pages:147b – 157b

S. Laxminarayan, B.H. Stamm, N. Piland, S. Weeg (2004), *Store & Forward Technology in Telemedicine*, In: Telemedicine Journal and e-Health, Suppl 1, S43

L. Bos, S. Laxminarayan, A. Marsh, (2005), *ICMCC: The Information Paradigm*, In: Medical and Care Compunetics 2, IOS Press, 1–4

B. M. O'Connell, S. Laxminarayan, (2005), *Understanding the Social Implications of ICT in Medicine and Health: The Role of Professional Societies*, In: Medical and Care Compunetics 2, IOS Press, 5–7

J. S. Suri, A. Dowling, S. Laxminarayan, S. Singh, (2005), *Economic Impact of Telemedicine: A Survey*, In: Medical and Care Compunetics 2, IOS Press, 140–156

R. Oberleitner, R. Wurtz, M. L. Popovich, R. Fiedler, T. Moncher, S. Laxminarayan, U. Reischl, (2005), *Health Informatics: A Roadmap for Autism Knowledge Sharing*, In: Medical and Care Compunetics 2, IOS Press, 321–326

R. S. H. Istepanian, C. S. Pattichis, S. Laxminarayn, (2006), *Ubiquitous m-health Systems and the Convergence towards 4G Mobile Technologies*, In: M-Health: Emerging Mobile Health Systems, Springer, 3–14

Guo Y, Sivaramakrishna R, Lu CC, Suri JS, Laxminarayan S. (2006). Breast image registration techniques: a survey. Med Biol Eng Comput. 2006 Mar;44(1–2):15–26

Proceedings, Books and Special Issues

R. Neubower. K. Foster, S. Laxminarayan (1987*)*, Co-Editors, *Proceedings of the 9th IEEE EMB International Conference on Biomedical Engineering*, Vol 9, Boston (2500 pages)

S. Laxminarayan (1988), Guest Editor, *Special Issue on Biomedical Supercomputing*, IEEE Engineering in Medicine and Biology Magazine, Supercomputing in Medical Imaging, Supercomputer Applications of Molecular Modeling, Supercomputer Applications in DNA Sequence Analysis, Large Scale Simulations of the Hippocampus, Supercomputer Applications in Orthopaedic Biomechanics Research, 7, 4, 11–43

J. Harris, C.J. Walker, S. Laxminarayan (1988), Co-Editors, *Proceedings of the 10th IEEE EMB International Conference on Biomedical Engineering*, Vol 10, New Orleans (2200 pages)

J. P. Morucci, R. Plonsey, J. L. Coatrieux, S. Laxminarayan (1992), Co-Editors, *Proceedings of the 14th IEEE EMB International Conference on Biomedical Engineering*, Vol 14, Paris, France (2900 pages)

J. P. Morucci, R. Plonsey, J. L. Coatrieux, S. Laxminarayan (1992), Co-Editors, *CD ROM Proceedings of the 14th IEEE EMB Conference on Biomedical Engineering*, Paris, France

P. Brett, S. Laxminarayan, *Proceedings of the International Workshop on Mechatronics in Surgery*, Bristol, United Kingdom, 1993

S. Laxminarayan, D. Kristol (1992), Guest Editors, *Special Issue on Computers in Medicine*, IEEE Engineering in Medicine and Biology Magazine, Challenges in the Human Genome Project, Exploiting Database Technology in Medicine, Integrated Imaging for Neuro Surgery, Voice-Driven Testing and Indtruction, Evoked Potential Research, and Electronic Environment for Biomedical Research and Education, 11, 1, 24–68

J. P. Morucci, R. Plonsey, J. L. Coatrieux, S. Laxminarayan (1992), Co-Editors, *International Symposium on Innovations in Biomedical Enginering in the Year of the European Unified Market*, IEEE Engineering in Medicine and Biology Society, Paris, France, Vol 14, 2769–2890

V. Thevenin, J. L. Coatrieux, S. Laxminarayan (1993), Guest Editors, *Special Issue on Biomedical Enginering Around the World*, Medical Technology Symposium of the European Economic Community, IEEE Engineering in

Medicine and Biology Magazine, The European Scene of Medical Technology, Biomedical Engineering Research in Europe, Biomedical Engineering: A View from the Industrial Stand Point, Biomedical Enginering Strategies, in the U.S., and Research Strategies in Japan, 12, 2, 42–63

S. Laxminarayan (1996), Editor, *Proceedings of the International Workshop on Internet: Past, Present and the Future*, NextGen Internet, USA, Chile, Argentina, Brazil, Mexico

S. Laxminarayan, E. Tzanakou (1998), Co-Editors, *Proceedings of the IEEE International Conference on Information Technology applications in Biomedicine*, Washington 12, Vol 2, IEEE Press

A. Marsh, S. Laxminarayan (1999), Co-Editors, *Proceedings of the IEEE International Conference on Information Technology Applications in Biomedicine*, Amsterdam, The Netherlands, Vol 3, IEEE Press

S. Laxminarayan (2000), Editor, *Proceedings of the IEEE International Conference on Information Technology Applications in Biomedicine*, Arlington, Virginia, Vol 4, IEEE Press

J. S. Suri, S. Laxminarayan, Book (2002), *PDE and Level Sets Algorithmic Approaches to Static and Motion Imagery Segmentation*, Springer Verlag

S. Laxminarayan, L. Kun (2002), Guest Co-Editor, *Special Issue on Bioterrorism*, IEEE Engineering in Medicine and Biology Magazine, Vol 21, No. 5

J. S. Suri, S. Laxminarayan, Book (2003), *Angiography Imaging: State-of-the-Art Acquisition, Image Processing and Applications Using Magnetic Resonance, Computer Tomography, Ultrasound and X-rays*, CRC Press Publications

S. Laxminarayan, L. Kun (2004), *Special Issue on Homeland Security*, IEEE Enginering in Medicine and Biology Magazine, Vol 23, No 1, 19–193

L. Bos, S. Laxminarayan, A. Marsh, Editors (2004), *Medical and Care Compunetics 1*, Studies in Health Technology and Informatics 103, IOSPress

J. S. Suri, D. Wilson, S. Laxminarayan, Editors (2005), *Handbook of Biomedical Image Analysis*, 3 volumes, Springer Verlag

J.S. Suri, C. Yuan, D.L. Wilson and S. Laxminarayan, Editors (2005), *Plaque Imaging: Pixel to Molecular Level*, Studies in Health Technology and Informatics 113, IOSPress

L. Bos, S. Laxminarayan, A. Marsh, Editors (2005), *Medical and Care Compunetics 2*, Studies in Health Technology and Informatics 114, IOSPress

R. Istepanian, S. Laxminarayan, C. Pattichi, Editors (2006), *Emerging Mobile E-Health Systems; Applications and Future Paradigm*, Springer Verlag

United States Congressional Hearings

Testimony on Telehealth and the Senior Health Mobile at the Senate Special Committee on Aging Field Hearing "High Tech Health Care" Reaching Out to Seniors through Technology" . Idaho State University, Pocatello, Idaho, July 12, 2002

Telehealth & E-Health Demonstration. Steering Committee on Telehealth And Healthcare Informatics. June 12, 2002, Capitol Hill, Dirksen Senate Office Building.

Associate Editor Columns (Sample Titles)

Fourth Young Researchers Forum, IEEE Engineering in Medicine and Biology Magazine, Vol 7, December 1998

EMBS Expands International Activities, IEEE Engineering in Medicine and Biology Magazine, Vol 11, 1, 20–22, March 1992

Development Proposal: International Institute of Human Organism Stability, IEEE Engineering in Medicine and Biology Society Magazine, Vol 11, 2, 91, June 1992

Biomedical Engineering Activities in South America: The First National Forum on Science and Technology in Health Care, IEEE Enginering in Medicine and Biology Magazine, Vol 12, 2, 89–90, June 1993

Health Policy in Brazil and its Influence on Clinical Engineering, IEEE Engineering in Medicine and Biology Magazine, Vol 12, 3, 22–23, September 1993

Rio de Janeiro Prepares to Host World Congress, IEEE Engineering in Medicine and Biology Magazine, Vol 13, 1, 19–21, March 1994

IEEE Hosts First India Regional Conference, IEEE Engineering in Medicine and Biology Magazine, Vol 14, 5, 648–649

UK Workshop Pinpoints Key Issues for Mechatronics in Surgery, IEEE Engineering in Medicine and Biology Magazine, Vol 15, May 1996

Information Technology Conference in Prague, IEEE Engineering in Medicine and Biology Magazine, Vol 17, January 1998

Colombian Bioengineering and Rehabilitation Seminar and Workshop, IEEE Engineering in Medicine and Biology Magazine, Vol 17, August 1998

Annual Congress of the Spanish Society for Biomedical Engineering, IEEE Engineering in Medicine and Biology Magazine, Vol 18, February 1999

Medicon'98 Held in Lemesos, Cyprus, IEEE Engineering in Medicine and Biology Magazine, Vol 18, April 1999

First Latin American Congress on Biomedical Engineering, IEEE Engineering in Medicine and Biology Magazine, Vol 18, March 1999

EMBS Goes to the Baltic Region, IEEE Engineering in Medicine and Biology Magazine, Vol 18, December 1999

Participation in IEEE USA CCIP Activities

- Opposing Adoption of the Uniform Computer Information Transactions Act, 2000

- Market Drivers: Competitive Analysis in Voice Over IP Technology, 2000
- Y2 K Liability, 1999
- Privacy On-Line, 1997
- Computer Crime, 1996
- Encryption Policy, 1996
- Spectrum (License) Auctioning, 1996
- High Definition Television Standards, 1997
- Privacy and Universal Identification Numbers,
- Digital Divide, *In Progress*
- Advanced Computing (Lead Author), *In Progress*

Technical Reports

S. Laxminarayan, M. Roederer (1965), ***Digital Flight Simulation Experiments: Flight Stability and Performance Simulations of the Airbus aeroplane***, Hamburger Flugzeugbau, Hamburg, West Germany

S. Laxminarayan (1965), ***Boundary Layer Theory***, Tech Rep 1.1, Aerodynamics Division, Hamburger Flugzeugbau, Hamburg, Germany

S. Laxminarayan, M, Zimmermann (1966), ***Aeroelasticity and Flutter Evaluations of the HFB320 Executive Jet***, Hamburger Flugzeugbau, Hamburg, West Germany

S. Laxminarayan (1966), ***Acoustic environment simulation using cyclic chain code siren and other facilities***, Tech Rep ISVR-NASA, University of Southampton, UK

S. Laxminarayan (1969), ***Data analysis using recursive and non-recursive digital filters***, DAC Tech Note, 1, ISVR, University of Southampton, UK

S. Laxminarayan (1969), ***Implementation of the Fast Fourier Transform Algorithm in machine code***, DAC Tech Note 4, ISVR, University of Southampton, UK

S. Laxminarayan (1969), ***Online digital data analysis system***, DAC Tech Rep, 2, ISVR, University of Southampton, UK

S. Laxminarayan (1970), ***Computerized archival system of analog and digital electrocardiograms for cardiac research***, Tech Rep, Thorax Center, Erasmus University, Rotterdam, The Netherlands

S. Laxminarayan (1971), ***A digital computer model of the excitation wave propagation in the human heart***, Tech Rep, Thorax Center, Erasmus University, Rotterdam, The Netherlands

S. Laxminarayan (1973), ***Signals processing methodologies in physiology***, Tech Rep, Free University, Amsterdam, The Netherlands

S. Laxminarayan (1980), ***Sudden Infant Death Syndrome patient database documentation***, Tech Rep, NICHD Contract, 1-HD-2555, Montefiore Hospital and Medical Center, New York

S. Laxminarayan, L. Michelson et al (1983), *Computer applications in biomedical research*, In: LOGGED ON, UMDNJ Publication, Newark, New Jersey

S. Laxminarayan (1987), *Editor, Annual Research Report*, Laboratory of Biomedical Engineering, UMDNJ, Newark, New Jersey

S. Laxminarayan (1989), *Guide to GenBank and DNA Sequence Analysis*, Technical Report, Academic Computing Center, University of Medicine and Dentistry of New Jersey

S. Laxminarayan (1990), *Molecular Modeling and Dynamics Applications Guide on HP Workstations*, Academic Computing Center, University of Medicine and Dentistry of New Jersey

S. Laxminarayan (1994), *UMDNJ Campus Wide Information System*, UMDNJ Publication, Newark, New Jersey

M. Majidi, D. Yatcilla, S. Laxminarayan (1995), *UMDWeb: University's gateway to Internet*, UMDNJ Publication, Newark, New Jersey

S. Laxminarayan (1995), *Internet and Internetworking Educational Protocols*, Tech Rep, John von Neumann Computer Network, Princeton

S. Laxminarayan (1999), *Voice and Video over Internet Protocol: Applications in Medicine and Health Care*, VocalTec White Paper

S. Laxminarayan (1999), *The Web Enabled Call Center*, VocalTec Technology Analysis Report

Presentations and Symposia (Sample List)

S. Laxminarayan (1966), *Jet Noise Simulation Using Psuedo Random Cyclic Chain Codes*, Institute of Sound and Vibration Research, University of Southampton, UK

S. Laxminarayan (1969), *Design of recursive and non-recursive digital filters*, Department of Electrical and Electronics Engineering, University of Southampton, UK

S. Laxminarayan (1970), *Spectral analysis of electrocardiograms for characterization of cardiac abnormalities*, Royal College of Surgeons, London, UK

S. Laxminarayan (1970), *Use of Graphics in Medical Data Visualization*, Royal College of Surgeons, London, UK

S. Laxminarayan (1971), *Digital Computer Modeling of the Excitation Wave Propogation in the Human Heart*, Thorax Center, Erasmus University, Rotterdam, The Netherlands

S. Laxminarayan (1980), *Arterial input impedance*, Biomedical Engineering Institute, Drexel University, Philadelphia, USA

S. Laxminarayan (1981), *Signals Processing Methodologies in Pulmonary Systems Characterization*, Albert Einstein College of Medicine, Bronx, NY, USA

S. Laxminarayan (1982), ***Biomedical Signals Processing***, Biomedical Engineering Department, New Jersey Institute of Technology, Newark, New Jersey, USA

S. Laxminarayan (1983), ***Computer applications in medicine***, Department of Anesthesia, College of Medicine and Dentistry, Newark, New Jersey, USA

S. Laxminarayan (1983), ***Sleep Profile from EEG Recordings in SIDS infants***, Indian Institute of Science, Bangalore, India

S. Laxminarayan (1984), ***Biomedical Signals Processing***, New York Academy of Sciences, Instrumentation Branch, New York, USA

S. Laxminarayan (1984), ***Sudden Infant Death Syndrome and Technology***, Department of Pediatrics, University of Medicine and Dentistry of New Jersey, Newark, NJ, USA

S. Laxminarayan (1984), ***Sudden Infant Death Syndrome and Technology***, United Childrens Hospital, UMDNJ Affiliate, Newark, New Jersey, USA

S. Laxminarayan (1985), ***Biomedical signals processing in a distributed environment***, Indian Institute of Science, Bangalore, India

S. Laxminarayan (1988), ***Genetic Engineering: Computational Resources, Tools, Functionalities and Limitations***, University of Medicine and Dentistry of New Jersey, Newark

S. Laxminarayan (1989), ***Spectral analysis of medical data***, Special Symposium on "State-of-the-art, Integrated, PC Based Software Systems", ASYST Software Technologies, Inc, Newark, NJ, USA

S. Laxminarayan (1989), ***Anatomy of Networking in Medical Environment***, New York Academy of Sciences, New York

S. Laxminarayan (1989), ***High Performance Computing Needs in Protein Engineering***, Robert Wood Johnson Medical School, Piscataway, New Jersey

S. Laxminarayan (1990), ***Biomedical Supercomputing***, IEEE Engineering in Medicine and Biology Society Chapter Meeting, Montreal, Canada

S. Laxminarayan (1989), ***Multi-institutional Networked Molecular Modeling: Cost-Effectiveness in Biomedical Education***, Stevens Institute of Technology, Hoboken (1989)

S. Laxminarayan (1993), ***Advanced computing technologies in medicine and biology***, Indian Institute of Technology, New Delhi, India

S. Laxminarayan (1993), ***Networking in biomedical research and education***, The John von Neumann Computer Network Symposium, Princeton, New Jersey

S. Laxminarayan (1994), ***Healthcare and Medicine on the Internet***, Global Enterprise Services, John Von Neumann Network, Argentina

S. Laxminarayan (1994), ***Computing and information technology in medicine***, IEEE EMBS, Distinguished lecture, New York, USA

S. Laxminarayan (1994), ***Computational Challenges in the Human Genome Project***, New York Academy of Sciences, New York, USA

S. Laxminarayan (1994), ***Molecular Modeling Initiatives in Drug Design***, School of Health Related Professions, UMDNJ, Newark, NJ

S. Laxminarayan (1994), ***Internet Medicine***, Distinguished Lecture, IEEE EMBS Chapter, NY

S. Laxminarayan (1995), ***Biomedical Engineers and Computational Biology***, Indian Institute of Technology, New Delhi, India

S. Laxminarayan (1995), ***Future of Radiology in the Information Age: Teleradiology, PACS and HIS***, University Hospital, UMDNJ. Newark, NJ

S. Laxminarayan (1997), ***Biomedical Information Technology: Quo vadis***, Distinguished Lecture, Faculty of Medicine, University of Sevilla, Spain

S. Laxminarayan (1997), ***Biomedical Information Technology: Health Care and Medicine***, Distinguished Lecture, Faculty of Medicine, University of Sevilla, Spain

S. Laxminarayan (2000), **Internet Medicine Revisited**, Distinguished Lecture, Tsinghua University, Beijing, China

Journal Citations

IEEE Transactions on Biomedical Engineering
IEEE Engineering in Medicine and Biology Magazine
IEEE Proceedings
IEEE Transactions on Information Technology in Biomedicine
Cardiovascular Research
Annals of Biomedical Engineering
American Journal of Physiology
Journal of Sleep Research
Circulation Research
Journal of Medical and Biological Engineering and Computing
Critical Reviews in Biomedical Engineering
Journal of Medical Engineering and Technology
Technology and Health Care
Journal of Pediatrics
Journal of Biomechanical Engineering
Journal of Physiology
Journal of Neuroscience
Journal of Hypertension
Methods of Information in Medicine
Journal of Applied Physiology
Bulletin of Mathematical Biology
Physiology Reviews
Biomaterials and Medical Devices

MIX
Papier aus verantwortungsvollen Quellen
Paper from responsible sources
FSC® C105338
FSC
www.fsc.org